IMPACT OF CALCULATORS ON MATHEMATICS INSTRUCTION

IMPACT OF CALCULATORS ON MATHEMATICS INSTRUCTION

Edited by
George W. Bright
Hersholt C. Waxman
Susan E. Williams

UNIVERSITY
PRESS OF
AMERICA

Lanham • New York • London

Copyright © 1994 by
University Press of America,® Inc.
4720 Boston Way
Lanham, Maryland 20706

3 Henrietta Street
London WC2E 8LU England

All rights reserved
Printed in the United States of America
British Cataloging in Publication Information Available

Library of Congress Cataloging-in-Publication Data

Impact of calculators on mathematics instruction / George W. Bright,
  Hersholt C. Waxman, Susan E. Williams, editors.
       p.   cm.
  Proceedings of a conference held in Houston, Tex., in May 1992.
  Includes bibliographical references.
  1. Mathematics—Study and teaching—Congresses. 2. Calculators—
  Congresses. I. Bright, George W. II. Waxman, Hersholt C.
       III. Williams, Susan E. (Susan Elaine).
  QA20.C34I47   1993   510'.71—dc20   93-33133 CIP

ISBN: 978-0-8191-9309-4

Acknowledgement and Disclaimer: The preparation of this volume was supported in part by a grant (Grant #R168D00311) from the U.S. Department of Education. All opinions expressed, however, are those of the authors and do not necessarily represent the positions of any government agency.

 The paper used in this publication meets the minimum requirements of American National Standard for Information Sciences—Permanence of Paper for Printed Library Materials, ANSI Z39.48-1984.

Contents

**Multiple Perspectives on the Impact of
Calculators on the Mathematics Curriculum**   1
    George W. Bright, Hersholt C. Waxman, & Susan E. Williams

Assessment and Technology   7
    Thomas A. Romberg

Calculator Inservice for Mathematics Teachers   27
    George W. Bright

**Using the Calculator to Develop Number and
Operation Sense, K - Grade 5**   41
    Jane F. Schielack & Clarence J. Dockweiler

**The Calculator Project: Assessing School-Wide
Impact of Calculator Integration**   49
    Gary G. Bitter & Mary M. Hatfield

**Research on Calculator Use in Middle School
Mathematics Classrooms**   67
    Juanita V. Copley, Susan E. Williams,
    Shwu-Yong L. Huang, & Hersholt C. Waxman

**The Graphing Calculator in Pre-Algebra Courses:
Research and Practice**   79
    Paul A. Kennedy

**The Calculator and Computer Precalculus Project (C^2PC):
What Have We Learned in Ten Years?**   91
    Bert K. Waits & Franklin Demana

Supercalculators in Undergraduate Mathematics   111
    Donald R. LaTorre

**Teaching Mathematics with Calculators (TMC):
A Local and National Inservice Teacher Education Project**   119
    John G. Harvey

**Future Directions for the Study of Calculators in
Mathematics Classrooms**   131
    Hersholt C. Waxman, Susan E. Williams, & George W. Bright

Multiple Perspectives on the Impact of Calculators on the Mathematics Curriculum

George W. Bright
University of North Carolina at Greensboro

Hersholt C. Waxman
Susan E. Williams
University of Houston

Many professional organizations in education have encouraged mathematics educators to reform the curriculum and incorporate more technology into the classroom instruction and activities of K-12 mathematics classrooms. The *Curriculum and Evaluation Standards for School Mathematics* (NCTM, 1989), for example, emphasize problem solving, higher-level thinking, and communication, *and* provide criteria for incorporating calculators into the school mathematics curriculum. The *Professional Standards for Teaching Mathematics* (NCTM, 1991) also advance the vision of high-quality instruction through the use of technology. Similarly, the Mathematical Association of America (MAA, 1991), Mathematical Sciences Education Board (MSEB, 1991), American Association for the Advancement of Science (Blackwell & Henkin, 1989), National Research Council (NRC, 1989, 1990), and Association for Supervision and Curriculum Development (Brandt, 1988; Willoughby, 1990) have all recommended the implementation of technology in all grade levels. These national calls for change, however, are probably not enough to assure the realization of wide-spread technology use in classrooms. Educators also need information on *why* it is important to integrate technology into the curriculum and *how* to integrate it effectively. This volume tries to address these two critical issues from multiple perspectives.

The chapters in this volume constitute the proceedings of the conference, "Impact of Calculators on Mathematics Instruction," held in Houston, Texas, in May 1992. The conference was organized at the request of the U. S. Department of Education as part of the second year work for the project, "Calculator Curriculum for 6-8 Mathematics" (Grant # R168D00311). The purpose of the conference was to provide a forum for reflection on how calculators are or might influence mathematics instruction. It was intended that recommendations might be developed at the conference that would become influential in directing future efforts on the integration of calculators into instruction.

Participants for the conference were selected to reflect most of the potential uses and multiple perspectives of calculators in mathematics instruction. We specifically sought speakers for the full range of K-12 mathematics, and we feel fortunate to have also been able to invite participants from the ranks of university mathematics professors. Clearly the expectations about what students will do with calculator power differ across grades, so readers should be alert to the different techniques that are suggested for engaging students in mathematics through calculator activities. We expected that the interaction arising out of the multiple perspectives of the participants would be exciting and enlightening. We were not disappointed, and we hope that readers of these papers will likewise be deeply stimulated.

Overview of the Book

This book is organized into two major sections. The first section focuses on issues related to calculator use. In the second section, several successful projects are highlighted and discussed. Finally, in the conclusion chapter we discuss directions for the study of calculators in the mathematics classroom.

The authors of several of the chapters in this volume use models that were developed in non-calculator settings to help them understand how calculators affect learning or how teachers can be helped to learn how to use calculators in instruction. Given the state of our understanding of calculator use in mathematics instruction, this type of "borrowing" is probably well advised and effective for our own sense making. However, calculators as an "innovation" may not be the same type of innovation as those which generated the models which are being borrowed. Calculator technology is rapidly changing, so what we learn about one type of calculator may or may not actually apply to newer types of calculators. It may ultimately be necessary for researchers to develop models that are specific to understanding changing technologies, even though we are not yet at the stage of being able to do this. Readers should be alert, therefore, for the use of these models and then should consider how the models might be modified to fit calculator environments better.

If calculators are to be integrated successfully into mathematics instruction, it is probably necessary to be sure that parents understand why and how calculators are being used. Parental attitudes often reflect a general societal bias against the use of calculators prior to students' mastering of paper-and-pencil techniques, especially at the primary level. It will take some effort on the part of teachers to convince parents that calculator use can actually assist students in understanding mathematical concepts. But in order for teachers to successfully argue this point with parents, it is probably necessary for teachers also to believe that calculator use does not have to wait until mastery of paper-and-pencil techniques. Inservice leaders have to be sure that teachers understand a variety of ways to use calculators for concept development. Many of the chapters in this volume address this point.

Issues Related to Calculators

In the first chapter of this section, "Assessment and Technology," Tom Romberg begins examining a series of questions about the relationship between assessment practices and the current reform movement in the mathematics curriculum. He addresses several key questions: (a) What does it mean to assess? (b) Why is assessment important in education? (c) Why are current assessment practices inadequate? (d) What new practices are being proposed? (e) What are some key issues related to the development of a new assessment system? and (f) What are the curricular implications of changes in assessments? In the next section, he summarizes the ways technological or cognitive tools can be incorporated in and expand upon the curriculum. In the final section, he illustrates the problem of creating "technologically-active" assessment tasks.

In his chapter, "Calculator Inservice for Mathematics Teachers," George Bright addresses several different areas of concern that calculator inservice programs should address. He first describes standard or low-level concerns: (a) calculator skill, (b) understanding of relevant mathematics, and (c) knowledge of pedagogy specific to the incorporation of calculators in instruction. He then describes more sophisticated or high-level concerns: (a) interaction between calculator use and testing and (b) teachers' beliefs about the role of calculators in learning mathematics. He concludes by discussing the importance of networking among teachers for the successful implementation of an innovation and by describing several approaches that might be taken to help teachers implement calculator-based instruction.

In their chapter, "Using the Calculator to Develop Number and Operation Sense, K-Grade 5," Jane Schielack and Clarence Dockweiler focus on the need to provide children with activities that incorporate the calculator as a tool to encourage thinking about mathematics by exposing children to patterns and relationships. They first consider the goals for the use of calculators in the mathematics classroom and then summarize the current research on reform in mathematics education especially as it relates to the constructivist approach to teaching mathematics. They then describe several calculator activities that can enhance concept development in young learners. Schielack and Dockweiler conclude by arguing that we need more knowledge about the ways children learn about mathematics with calculators.

Successful Calculator Projects

The chapter by Gary Bitter and Mary Hatfield, "The Calculator Project: Assessing School-Wide Impact of Calculator Integration," describes a study that primarily examined the effect of calculator use on middle school students' standardized test scores in mathematics. They found that eighth-grade students improved on all three Iowa Test of Basic Skills mathematics subtests, while seventh-graders significantly improved on the computation subtest. They also found that students reported using calculators more during higher-order activities than for computing grades and checking work. Parents' and teachers' attitudes toward calculators also improved from the beginning to the end of the project.

In the next chapter of this section, Copley, Williams, Huang, and Waxman describe research that was conducted during the first year of a U. S. Department of Education, Dwight D. Eisenhower Mathematics and Science Program grant which investigated calculator use and calculator curriculum development in middle school mathematics classrooms. The descriptive, observational findings revealed that calculators were primarily used for computation-focused activities and for verifying answers. The results from the teacher survey revealed that teachers were generally positive toward calculators, but less than half indicated that students should be allowed to use calculators while taking mathematics tests. The quasi-experimental results revealed several significant differences between the experimental and control groups on calculator use in the classroom. Experimental teachers were observed significantly more often than control teachers (a) allowing students to determine the appropriate use of the calculator, (b) emphasizing the importance of estimation for determining the reasonableness of a calculator answer, and (c) stressing the use of a calculator as a problem-solving tool.

Paul Kennedy's chapter, "The Graphing Calculator in Pre-Algebra Courses: Research and Practice," highlights a project funded by the U. S. Department of

Education: Partnership for Access to Higher Mathematics (PATH Mathematics). PATH Mathematics is a program for eighth- and ninth-grade at-risk students and provides them access to Algebra I and other advanced mathematics courses. The program uses the graphing calculator throughout the curriculum as a tool for computation, concept development, and problem solving. Some of the preliminary research results from the pilot semester show that students have positive attitudes toward calculator use, but they also still feel that using a calculator sometimes seems like cheating. He also found that positive attitudes toward mathematics and calculator use have a positive impact on students' performance in algebra.

In their chapter, "The Calculator and Computer Pre Calculus Project (C^2PC): What Have We Learned in Ten Years?," Bert Waits and Frank Demana describe the philosophy and background of the C^2PC project which was designed to increase the number of students ready for university-level calculus. They also illustrate how the hand-held visualization technology of the graphing calculator is incorporated in 10 types of fundamental activities that occurred every day in their project classrooms. They conclude by advocating new assessment methods and additional studies on what C^2PC is doing and what C^2PC students are learning.

In his chapter, "Supercalculators in Undergraduate Mathematics," Don LaTorre describes a Clemson University project that effectively integrated HP-48S/SX supercalculators into the calculus, differential equations, and linear algebra sequence. He describes the impact the supercalculator has had on students and faculty and in particular highlights the student perception data that indicates that students generally felt that supercalculators (a) helped them understand the materials, (b) allowed them to do more exploration and investigation in solving problems, and (c) helped them have a better intuition about the material. La Torre concludes his chapter by highlighting the advantages of high-level supercalculators and maintaining that the project had a strong and long-lasting impact.

In the last chapter of this section, "Teaching Mathematics with Calculators (TMC): A Local and National Inservice Teacher Education Project," John Harvey presents a rationale for national inservice on calculator use. He then describes the local inservice efforts that supported the development of materials that have been, and continue to be, disseminated nationally as a means of promoting effective use of different calculators in a variety of mathematics instructional settings. The cooperation evident among funding agencies, professional organizations, school districts, and industry is a model that others may wish to examine closely.

Conclusions

In the concluding chapter of this book, we describe four specific areas that have important implications for future research in the field. First, we examine the impact of theory and research from the field of cognitive psychology on the use of calculators in mathematics classrooms. Second, we discuss needed research studies in the area of staff development and training. Third, we briefly look at issues related to evaluative research studies. Finally, we discuss the need to have more programmatic research in this area.

Calculators can provide teachers and students a chance to view mathematics differently by having the power available to handle "messy" computations such as are found in real world problems and to explore patterns through the generation of data that would be too time consuming to generate by hand. The chapters in this volume suggest that teachers are enthusiastic toward calculators, but we also know that they need to be retrained in order to incorporate calculators effectively into the mathematics curriculum. Inservice education on calculators, however, is not as easy as it may appear. Effective inservice programs have to focus on both affective and cognitive components. The mathematics curriculum may also need to be modified or restructured so that teachers can effectively implement calculators in the curriculum.

The changing context dictated by changes in calculator technologies may also cause us to change our views of appropriate uses of calculators. We hope, however, that general principles about calculator use can be developed so that we will have a framework within which to understand the effects of changing technology.

There are, of course, many unanswered questions about the effects on students of using calculators over the long term. Although some of the chapters in this volume provide some evidence for the effectiveness of calculators on students' cognitive and affective outcomes, more research is obviously needed in this area. We had hoped that this present volume would be able to provide a synthesis both of what we think we know and what we need to know relative to these concerns. We were probably less successful in developing this synthesis, not because of the quality of the presentations but because of the lack of information that can be brought to bear. As we will point out in our concluding chapter, there is much work to be done. We hope that these papers will encourage others to join us in the search to begin to answer some of the interesting questions.

We want to thank the U.S. Department of Education, the University of Houston, and Texas Instruments for their generous support of the conference. We believe that all the participants gained knowledge as well as a renewal of energy and enthusiasm about the potential benefits of using calculators in mathematics instruction. We also gained a sense of direction about what questions need to be answered next, and we hope that those who read these papers will gain a similar understanding.

We want to thank all the participants of the conference for sharing the adventure with us. Their multiple perspectives truly enriched all our understandings of the complexities of integrating calculators into the mathematics curriculum. In addition to the presenters, we would like to thank the chairs of the conference sessions (Joe Dan Austin, Rice University; Pam Chandler, Fort Bend Independent School District; Greg Foley, Sam Houston State University; Stephanie Knight, University of Houston; Kathryn Lay, Grambling State University; Bonnie McNemar, Harris County Department of Education; Emiel Owens, Prairie View A & M University; Michele Rohr, Houston Independent School District; and Patti Wooten, Aldine Independent School District) for facilitating the dialogue that occurred. We also appreciate the classroom teachers (Fae Knight, Jan Moore, and Laurye Webb) and mathematics supervisor (Marsh Lilly) from Alief Independent School District in Alief, Texas, who shared their perspectives with us. But most of all, we want to thank the authors who gave generously of their time to write and revise the papers in this volume.

References

Blackwell, D., & Henkin, L. (1989). *Mathematics: Report of the Project 2061 Phase I Mathematics Panel.* Washington, DC: American Association for the Advancement of Science.

Brandt, R. S. (Ed.). (1988). *Content of the curriculum.* Alexandria, VA: Association for Supervision and Curriculum Development.

Mathematical Association of America. (1991). *A call for change: Recommendations for the mathematical preparation of teachers of mathematics.* Washington, DC: Author.

Mathematics Sciences Education Board. (1991). *Counting on you: Actions supporting mathematics teaching standards.* Washington, DC: National Academy Press.

National Council of Teachers of Mathematics. (1989). *Curriculum and evaluation standards for school mathematics.* Reston, VA: Author.

National Council of Teachers of Mathematics. (1991). *Professional standards for teaching mathematics.* Reston, VA: Author.

National Research Council. (1989). *Everybody counts: A report to the nation on the future of mathematics education.* Washington, DC: National Academy Press.

National Research Council. (1990). *Reshaping school mathematics: A philosophy and framework for curriculum.* Washington, DC: National Academy Press.

Willoughby, S. S. (1990). *Mathematics education for a changing world.* Alexandria, VA: Association for Supervision and Curriculum Development.

Assessment and Technology

Thomas A. Romberg
University of Wisconsin - Madison

In the recent literature on mathematics education, there is a consistent call for new procedures for assessing mathematics achievement (e.g., Campbell & Fey, 1988; Jones, 1988; National Research Council, 1989; National Council of Teachers of Mathematics [NCTM], 1989; Oakes, 1986; Romberg, 1988). The rationale for this call, as summarized by Webb (1987), is as follows. The increasing importance of mathematics in society and a new understanding of how children learn have led to demands for reform in the mathematics curriculum. It is thought that current assessment procedures, because they are based on earlier views about mathematics and learning, will inhibit curricular reform. Advocates of new assessment procedures argue that new methods must be developed that better reflect not only current understanding about how knowledge is constructed but also the mathematics that students should learn. In short, "as the curriculum changes, so must the tests" (NCTM, 1989, p. 214).

My purpose in this paper is threefold. First, I examine a series of questions about assessment practices in relationship to the reform movement. Second, I discuss the role of technological tools in relationship to new assessment practices. Third, I illustrate the problem of creating "technologically-active" assessment tasks.

Questions about Assessment

What does it mean to assess? Bodin (1991) argues that "to assess means to organize (or to look at) a situation in such a way that it enables us to gather information which, after processing, can reveal something that is reliable about personal knowledge (or about the collective knowledge of a group)" (p. 4). Thus, the main concern of assessment should be to gather reliable information about individual or collective knowledge. This definition implies that any assessment is based on assumptions about the nature of knowledge.

Why is assessment important in education? There are two general answers to this question. The practical answer is captured in the following cartoon (Figure 1). Assessment helps us distinguish between teaching and learning. Goodson (1988) calls the relationship exhibited in the cartoon "the trilogy of pedagogy, curriculum, and examinations" (p. 32). In practice, one cannot divorce assessment from content or how that content is taught.

The political answer is captured in the relationships between specifications of content, instruction, and assessment as shown in Figure 2 (Romberg & Wilson, in press). This figure illustrates the *America 2000* (U.S. Department of Education, 1991) strategy that has been adopted by the President and the National Governors' Association for the restructuring of American schools. The strategy involves a system in which: agreement is reached on a detailed set of content standards in

English, mathematics, science, history, and geography; agreement is reached about how best to instruct students toward those content standards; procedures are developed to accurately assess student progress in meeting those standards; and the professionals are held accountable for the results of their efforts to assist students in reaching those standards.

Figure 1. *It is assessment which helps us distinguish between teaching and learning.*

What is also shown in Figure 2 is the fact that the National Council on Education Standards and Testing (1992) has decided that the model for initiating this discipline-based reform effort is the approach taken by the mathematical sciences education community. In particular, the *Curriculum and Evaluation Standards for School Mathematics* (NCTM, 1989) presents a consensual vision of the mathematical content that all students should have an opportunity to learn--the **content standards** in Figure 2. Furthermore, the *Professional Standards for Teaching Mathematics* (NCTM, 1991) describes the appropriate means for assisting students to learn that content--the **instructional standards** in Figure 2. In addition, some of the needed elements for the development of an assessment system are described in the Evaluation Standards in the first document.

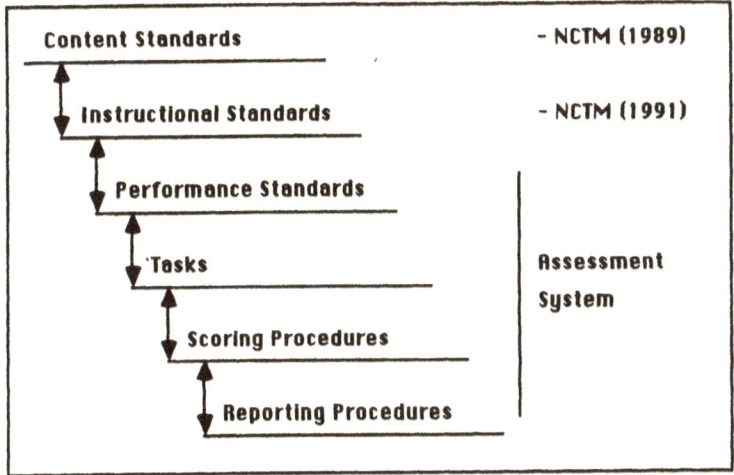

Figure 2. *Relationships between content, instructional and performance standards, and an assessment system.*

Why are current assessment practices inadequate? The answer to this question rests in an epistemological shift in our definition of what it means to know mathematics (Romberg, 1992). Most current mathematics assessments involve multiple-choice test items. Although critics claim that there are many problems with such assessments, there are two criticisms that seem to be most frequently raised. First, critics argue that the tests reflect a fragmented view of mathematics rather than the view of mathematics as an integrated whole, as described earlier in this report. According to Webb and Romberg (1988):

> Most multiple-choice or fixed-choice tests ... are designed to measure independent partitioning of mathematics rather than knowledge and use of the interrelationships among mathematical ideas. These tests are organized based on instructional ... competencies that reflect a view of mathematics as a large collection of separate skills and concepts. (p. 3)

The second frequently voiced criticism is that current mathematics tests predominantly consist of items requiring factual knowledge or procedural skills rather than items that assess students' abilities to think critically, to reason, to solve problems, to interpret, and to apply ideas in creative ways. Campbell and Fey (1988) assert that "current standardized assessments ... are promoting exposure to procedures at the expense of understanding" (p. 62). According to the National Research Council (1989), current "tests stress lower- rather than higher-order thinking, emphasizing student responses to test items rather than original thinking and expression" (p. 68). With their focus on solutions rather than on strategies or processes, the current tests "reinforce in students, teachers, and the public the narrow image of mathematics as a subject with unique correct answers" (National Research Council, 1989, p. 68).

Two assumptions underlie the new notions about "knowing mathematics."

Assumption 1. We are now in a new age--*The Information Age*--which will significantly alter the character of American schooling.

Some of the attributes of the shift from an industrial society to an information society are: (a) it is an economic reality, not merely an intellectual abstraction; (b) the pace of change will be accelerated by continued innovation in communications and computer technology; (c) new technologies will be applied to old industrial tasks at first, but will then generate new processes and products; and (d) basic communication skills are more important than ever before, necessitating a literacy-intensive society.

Zarinnia and Romberg (1987) argued as follows:
> The most important single attribute of the Information Age economy is that it represents a profound switch from physical energy to brain power as its driving force, and from concrete products to abstractions as its primary products. Instead of training all but a few citizens so that they will be able to function smoothly in the mechanical systems of factories, adults must be able to think. ... This is significantly different from the concept of an intellectual elite having responsibility for innovation while workers take care of production. (pp. 23-24)

In fact, since on average, most workers will change jobs four to five times, they can no longer assume they can acquire initially the mathematical skills needed throughout their years in the workplace. A flexible work force capable of lifelong learning is now required.

Assumption 2. Higher order thinking skills must be the focus of instruction in mathematics.

All learning involves thinking, but in the past, most instruction focused on learning to name concepts and follow specific procedures. Now the emphasis for all students must shift to communication and reasoning skills. Although these skills resist precise definitions, they are now popularly called "higher order" thinking skills. Resnick (1987) listed some of their features, many of which are in stark contrast to current mathematics criteria, which are shown in parentheses.
1. Higher order thinking is *nonalgorithmic*. That is, the path of action is not fully specified in advance. (Still largely algorithmic)
2. Higher order thinking tends to be complex. The total path is not "visible," mentally speaking, from any single vantage point. (Standard examples with visible paths)
3. Higher order thinking often yields *multiple solutions*, each with costs and benefits, rather than unique solutions. (Single unique solutions)
4. Higher order thinking involves *nuanced judgment* and interpretation. (Neither judgment nor interpretation expected)
5. Higher order thinking involves the application of *multiple criteria*, which sometimes conflict with one another. (Simplified to single criteria that are well-defined in content)
6. Higher order thinking often involves *uncertainty*. Not everything that bears on the task at hand is known. (Certain--all information required is given)

7. Higher order thinking involves *self-regulation* of the thinking process. We do not recognize higher order thinking in an individual when someone else "calls the plays" at every step. (External regulation)
8. Higher order thinking involves *imposing meaning*, finding structure in apparent disorder. (Meaning is given or assumed)
9. Higher order thinking is *effortful*. There is considerable mental work involved in the kinds of elaborations and judgments required. (Work which usually involves standard exercises is simplified so that little effort is needed) (pp. 2-3)

Given these contrasts, then, thinking skills must be the focus of instruction in mathematics in the near future, and assessment procedures need to be developed that portray not only the number of correct answers students can produce, but the thinking that produced those answers.

What new practices are being proposed? Critics of current mathematics assessment procedures are consistent in what they propose to replace the current procedures. Campbell and Fey (1988) argue that "it is essential that testing be reorganized to ... parallel the new goals of the elementary curriculum: the development of mathematical understanding, the interpretation of mathematical events, and the application of mathematical procedures" (p. 62). As cited earlier, Donovan and Romberg (1987) believe that the emphasis should shift from product to process. Similarly, Oakes (1986) cites a "critical need for better indicators" of educational achievement, and notes that "we have fairly good paper-and-pencil measures of the most commonly taught basic knowledge and skills. But we lack adequate measures of children's abilities to think critically, to apply their knowledge, or to solve problems" (p. 34).

The NCTM *Standards* (1989) provides more specific suggestions regarding what should be assessed. There are seven student assessment standards that focus on the assessment of students' understanding of and disposition toward mathematics. Each of these standards is briefly described below. It is suggested in the *Standards*, however, that the student assessment standards are most relevant to teachers and classroom assessment. Although these standards have not been specifically proposed for state-level assessment, many of their features may be applicable for this purpose. The seven student assessment standards, as described in the NCTM *Standards*, are as follows:
1. *Mathematical power.* This broad standard focuses on the integration of the abilities covered in the other assessment standards. It focuses on the extent to which students (a) have integrated the information they have learned, (b) can apply what they have learned to problem situations, (c) can communicate their ideas, (d) have confidence in doing mathematics, and (e) value mathematics.
2. *Problem solving.* Students should be assessed on their ability to use mathematics to solve problems. All aspects of problem solving should be covered.
3. *Communication.* Assessment of students' ability to communicate mathematically should focus on both the meanings students attach to concepts and procedures, and on their fluency in talking about, understanding, and evaluating ideas expressed through mathematics.

4. *Reasoning.* Assessment techniques should specifically assess students' ability to use various types of reasoning that are fundamental to mathematics (e.g., deductive, proportional).
5. *Mathematical concepts.* Assessments of students' knowledge should examine their understanding of mathematical concepts.
6. *Mathematical procedures.* Assessments of students' knowledge of procedures should determine not only whether students can execute procedures, but also whether they know the underlying concepts, when to apply the procedures, why the procedures work, and how to verify that the procedures yield correct answers.
7. *Mathematical disposition.* Assessments should seek information about students' attitudes toward mathematics, including confidence, willingness to explore alternatives, perseverance, and interest.

What are some issues related to the development of a new assessment system? Romberg and Wilson (in press) identified nine issues that need to be addressed if a new assessment system is to be developed. Their first two issues have already been discussed:

Issue 1. Underlying assumptions about the nature of mathematics
If one considers mathematics to be a static, linearly-ordered set of discrete facts, then the logical choice for a valid assessment system is the traditional standardized achievement test. On the other hand, if one views mathematics as a dynamic set of interconnected, humanly constructed ideas, then the assessment system must allow students to engage in rich activities that include problem solving, reasoning, communications, and making connections.

Issue 2. Underlying assumptions about the learning of mathematics
Assessment should be based on a view of the learning of mathematics as a socially-constructed process, not a fixed hierarchy of skills and concepts to be mastered.

Issue 3. The need for new psychometric models
"It is only a slight exaggeration to describe the test theory that dominates educational measurement today as the application of twentieth century statistics to nineteenth century psychology" (Mislevy, in press). Mislevy is calling for the field of psychometrics to "catch up" with the advances in cognitive psychology. Just as an examination system must be built on current learning theory, so must the psychometrics that support such a system be designed with cognitive psychology as its base.

Issue 4. Alignment with the reform curriculum
As described in Figure 2, the first stages of the building of an assessment system for mathematics, that is, the setting of content and instructional standards, has been accomplished. Consensus has been reached in the mathematics education community about the content that all students should be given the opportunity to learn and about the appropriate means of instruction. As the next five stages (setting performance standards, developing tasks, measuring student growth, adopting scoring criteria, and reporting procedures) are undertaken, it is critical that the outcomes are in alignment with the conceptualizations of curriculum and instruction set forth in the *Standards*.

Issue 5. Specification of performance standards

The NCTM *Curriculum and Evaluation Standards* describe what students should have an opportunity to learn. But in order to establish an examination system for school mathematics that is aligned with that vision, performance standards must be set that will describe what students are supposed to know and be able to do in mathematics. Making the connection between curriculum standards and performance standards is a difficult task, but one that needs to be done.

Issue 6. Developing authentic tasks

Increasing attention is being given to notions of "authentic assessment." Definitions of criteria for authentic assessment are being developed that are built on the framework of the reform curriculum in mathematics education. In order for an examination system to be considered "authentic," it must take cognizance of these criteria.

Archbald and Newmann (1988) consider three criteria to be critical to authentic assessment tasks: (a) disciplined inquiry, (b) integration of knowledge, and (c) value beyond evaluation. Disciplined inquiry refers to the production of new knowledge, such as that created by scientists or historians. It depends on prior conceptual and procedural knowledge, it develops in-depth understanding of a problem, and it "moves beyond knowledge that has been produced by others" (p. 2). Integration of knowledge means that authentic tasks must consider the content as a whole, rather than as a collection of knowledge fragments. Students must "be challenged to understand integrated forms of knowledge," and "be involved in the production, not simply the reproduction, of new knowledge, because this requires knowledge integration" (p. 3). The third criterion, value beyond evaluation, refers to the idea that authentic tasks should possess attributes that make them worthwhile activities beyond their use as evaluative tasks.

Issue 7. Measuring status, growth, or a combination

It is clear that all forms of assessment, including traditional standardized tests, measure the present status of student thinking. Traditional measuring instruments were built to yield highly reliable scores on a single dimension, with the ultimate purpose being to linearly rank students on that dimension. This factory-like image of education belongs to an earlier, Industrial Age. Constructivist approaches to assessment, suitable for the Information Age, require a greater emphasis on differentiated profiles of individual students. There is a need for more than status information; instead of a static score, what is needed is a measure of growth over time.

As Bodin (1991) argues, "a person's knowledge, fortunately for itself and unfortunately for its assessment, is not set once and for all. Knowledge essentially is changeable, subject to rearrangements, and more often than not, appears in disguised form" (p. 5). In Figure 3 judgments about students' knowledge is the key element in tracking the progress of the teaching-learning process. The boldest arrow at the bottom in Figure 3 reminds us that decisions about educational context are fed not only from knowledge-centered analysis. Numerous other deciding factors can interfere, factors that are linked with school policy, teachers' perceptions, etc., making it necessary to split off those factors in order to clarify the assessment process.

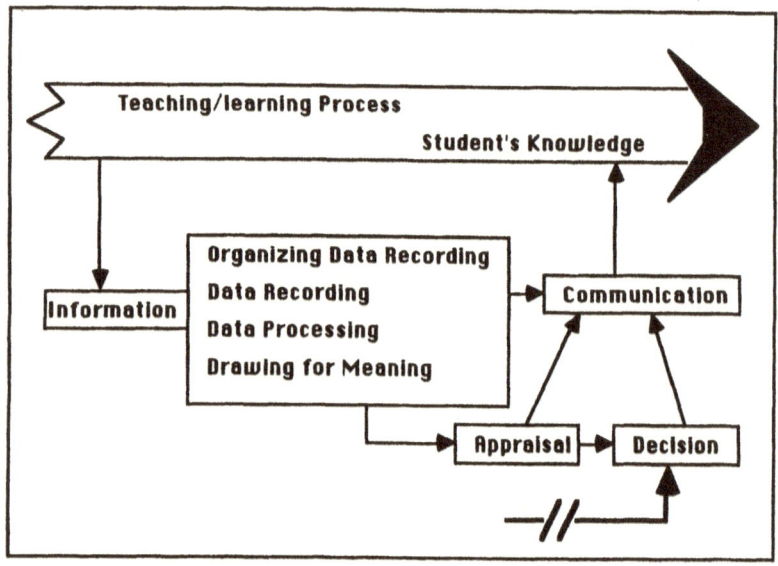

Figure 3. *Decision making.*

Issue 8. Scoring--by whom and in what form?
Scoring on traditional standardized tests has historically been done by machine, and that works fine with multiple-choice, single-answer items. But different forms of assessment with open-response items, for example, require professional judgment to score. The issue is, can we trust teachers to reliably mark their own students' work, and can they be trained to do so? This notion of the importance of teachers is a critical point, for as Stake (in press) has argued:

> In this last decade of the 20th Century, education remains labor-intensive. Efforts to automate teaching have been largely unsuccessful. Why do we continue to put at least one expensive laborer in each classroom of every school? It is not because teacher unions are featherbedding. It is not because it takes a scholar to maintain discipline. It is not primarily to assure the choice and presentation of subject matter. We place almost three million teachers in American classrooms because managing the conditions for learning requires constant attention--recognizing readiness to learn, the uniqueness of students, obstacles and intrusions, perceiving "progress within wrong answers," a never-ending assessment of student achievement. (p. 17)

Issue 9. Making reports of results understandable to the public
Results of student performance on an examination need to be reported to several audiences: students, parents, teachers, administrators, and policy makers. The form and substance of these reports will necessarily vary according to the audience. Nevertheless, it is essential that they are designed to be easily understandable to their constituents at the same time that they preserve the richness of the information. It would do little good to replace traditional tests with non-traditional formats if the

conceptually abundant information gathered is collapsed into a single numerical score or left in an uninterpreted form.

The first question that must be addressed in the design of reports is what kinds of information do students, parents, teachers, administrators, or policy makers need in order to make informed decisions? Since even the unit of analysis is different for each audience, the information required will likewise vary. Once that answer is decided, the appropriate data sources and means of analysis can be determined.

In summary, many mathematics educators are calling for new methods of assessing students' mathematical abilities. The call is for assessment methods that reflect both the current view of learning as a constructive process and the changes being recommended in the mathematics curriculum.

What are the curricular implications of changed assessments? The potential impact of testing on curriculum is widely recognized. Frederiksen (1984) describes the influence that testing can have on what is taught and learned.

> Any test is likely to influence the behavior of students and teachers, provided they know about it in advance. Students want to get respectable grades, or at least pass the course, and teachers want their students to do well. Thus, tests tend to increase the time and effort spent in learning and teaching what the tests measure and (to the extent that amount of time and effort is fixed) decrease efforts to learn and teach skills not measured by the test. (p. 193)

If the focus in the assessment of mathematics shifted from product to process, then conceptual understanding, problem solving, and conceptual development would probably receive more emphasis in the classroom. Furthermore, a change in the types of items used on mathematics tests could also impact mathematics curriculum. Frederiksen notes that item type "may influence the cognitive processes involved in dealing with test items and, hence, the nature of the skills taught and learned" (1984, p. 195). There is some empirical support for this idea. For example, a review of the effects on students of test format concluded that test format influenced the way in which students prepared for the test. It seems likely that these results would generalize to mathematics tests.

> When students expected an objective test, they studied for details and memorization of material. When studying for an essay test, they focused attention on relationships, trends, organization, and on identifying a few major points around which they could write descriptive or expository paragraphs. (Kirkland, 1971, pp. 315-316)

Tests send a message to both students and teachers about what it is important to learn. If mathematics tests were used that measured students' knowledge and understanding of concepts and procedures, their ability to reason and analyze, and their ability to apply their knowledge to solve problems, these skills would undoubtedly receive more attention in the mathematics curriculum.

In summary, the following remarks aptly summarize the argument put forth by advocates of new assessment procedures in mathematics.

> Although it may take years before current testing procedures are replaced in schools, we should be looking at plausible alternatives. Our

task is <u>not</u> to write cleverer test items of the same type (independent, single answer, multiple choice), but to write other kinds of questions based on this new conception of domain knowledge. We need questions that give students an opportunity to think like a mathematician--reflect, organize, model, represent, argue. In addition, these questions should be structured so that students' knowledge of the interrelationships between ideas within specific mathematical domains can be determined. Constructing, scoring, scaling, and interpreting responses to such items will not be easy, but in the long run, they will be worth the effort. (Romberg, 1988, pp. 32-33)

The Role of Technological Tools

What would a classroom look like, given the vision presented in the NCTM *Standards*? First, the atmosphere should be an equitable blend of order and disorder. As students work through problems, a great variety of tools should be made available so that students can try a number of different strategies, determine the "best" strategy for themselves, and understand why different strategies may be equally appropriate. To maintain commonality of purpose in this atmosphere which appears disorderly, communication between students, and between students and teachers, is a must. Students should be encouraged to create projects, write summaries, draw pictures, and discuss the reasonableness of answers.

Second, among the variety of tools students use to solve problems, some should be general tools applicable across topics and units (calculators, for example), and some should be specific to topics (statistics packages, for example). Some should be sophisticated, like electronic equipment, and some should be simple, but powerful, like graph paper. All tools should be appropriate for the age and mathematical knowledge of the students.

The notion of "technological tools" is based on the fact that even at its most advanced, interactive state, technology is merely a tool for teachers and students to explore possibilities. "Cognitive tools" as simple as a pencil, or a set of counters, free up human memory limitations so that students can focus on the more important aspects of mathematics, such as problem solving and communication. To understand the role of computers and calculators in mathematics assessment, it is useful to think about the role that technology is playing in the student's learning. Pea (1987) identifies two function types for what he calls "cognitive technologies: ... (a) those which promote PURPOSE--engaging students to think mathematically; and (b) those which promote PROCESS--aiding them once they do so" (p. 100). For most technologies, the function is clearly purpose or process, although for some, this distinction might be hard to make.

An example of a multimedia technology that is in the "purpose" category is the "Voyage of the Mimi" (Char, Hawkins, Wootten, Sheingold, & Roberts, 1983). Using video, software, and print materials, students follow the adventures of scientists aboard the *Mimi* as they search for and study whales. Through the experiences that are presented, students are engaged to solve problems that integrate concepts in mathematics, science, social studies, and language arts. Students see how mathematics is found in these realistic situations, and they are led to use mathematics in order to help them make sense of what they experience.

Pea's (1987) categorization of technologies that promote "process" is more complex. He makes a distinction among the following functions for tools: developing conceptual fluency, mathematical exploration, integrating different mathematical representations, learning how to learn, and learning problem-solving methods. Examples of tools for mathematical exploration and integrating different mathematical representations will illustrate how technologies can aid students in thinking mathematically. The *Geometric Supposer* series (Schwartz & Yerushalmy, 1985) on triangles, quadrilaterals, and circles enables students to collect visual and numerical data about figures in order to form and test conjectures. The software encourages exploration because an observation of what might be true in a single example can be easily tested in as many other examples as the student wants. This informal, inductive approach can be an effective lead-in to a more formal, deductive, proof-oriented course.

Representing concepts in multiple ways and being able to link those representations is being recognized as important in the learning process (Kaput, 1989). Multiple representations of algebraic functions can be explored with *The Function Analyzer* (Schwartz, Yerushalmy, & Education Development Center, 1988), which also allows students to see the interplay between the graphic, algebraic, and data table representations of functions. Changes in one representation are visibly mirrored by changes in the other two representations. Experimentation, in a way that is just not possible in a classroom with only a blackboard or overhead, now becomes commonplace.

It is important to note that graphics calculators allow the same multiple representations of functions. Students can enter the algebraic notation for multiple functions and the calculator will produce their graphs. Through a trace capability, the student can find the coordinate values of points on a graph. This feature enables students to find maximum and minimum values, points of intersection, and roots of functions. Graphics calculators such as the Texas Instruments TI-81 and TI-85 are easy to use, fast, and very appealing to students who might be otherwise intimidated by a computer. They also provide a price advantage over computers and software. As Kepner (1989) points out, for less than the cost of a pair of athletic shoes, a graphing calculator can be put into the hands of every secondary mathematics student 24 hours a day. This is mathematical and pedagogical power that we cannot ignore.

There are other tools for learning besides calculators and computers. We know the importance of having students use manipulatives when they are learning new mathematical concepts in the elementary grades. However, the potential of manipulatives in secondary school mathematics has been largely ignored. We usually skip any opportunity to have students model situations and instead move swiftly to a symbolic representation whenever possible.

In summary, the following global characteristics illustrate the ways in which cognitive tools can be incorporated in and expand upon the curriculum.
1. Tools can play an important role in improving the didactics of mathematics education.
2. Tools can be a strong exploration medium for students. Students can explore a new environment, working with their own strategies and at their own speed.

3. Tools can be a strong instructional aid for the teacher, used by the teacher when introducing a new topic, or shedding new light on old ideas.
4. Tools can improve the way in which some areas of mathematics can be taught, or even make it possible to enter more complex areas that could not be approached without the use of technology.
5. Tools can create new curricula, widening the boundaries of the current curriculum. In addition, the tools themselves can be a subject of exploration.
6. Tools are not only devices for exploring the "real world." The needs of business and industry dictate that computer tools *are* the real world, or at least a major factor for success in the real world.
7. Tools in the mathematics classroom mean changed roles for the teacher and student. Thus, attention must be paid to implementation. This means that both software and written curricula must be easy to use and written according to what we know about teacher/student interaction in the learning process.
8. Tools are an essential component of instruction and must be fully integrated into the learning materials.
9. Tools are not language-dependent. Manipulatives and calculators can be understood by *all* students. In addition, software, text, and video can be translated into any language. Thus, the potential for multicultural education is enhanced with appropriate tools.

Cognitive tools are not a substitute for good teaching, but rather represent a complementary component of good teaching. If used properly, computers, calculators, manipulatives, and other technologies can be indispensable instruments for facilitating a better understanding of mathematics (Corno & Snow, 1986; Edwards, 1991).

"Technologically-Active" Assessment Tasks

The key implication of the previous discussion about cognitive tools is that if they are used in instruction, they should be used in any assessment. In fact, the initial evaluation standard in the *Curriculum and Evaluation Standards* states that "methods and tasks for assessing students' learning should be aligned with the curriculum's ... instructional approaches and activities, including the use of calculators" (1989, p. 193).

Unfortunately, in spite of this emphasis, I have been unable to locate many examples of assessment tasks where students will need to use any technological tool. For tests in which "calculator-active" items are included, Harvey's (1992) review is most comprehensive. I will not repeat all of his arguments here. However, I have chosen to use his terms.

> Because there have been no recommendations about the ways in which tests should be changed, three approaches have been used that permit students to use calculators while taking tests. These approaches
> 1. permit students to use calculators, but give them tests that make no provision for calculator use. I will call this approach *calculator-passive testing*.
> 2. permit students to use calculators, but give them tests developed so that none of their items require calculator use. This approach will be called *calculator-neutral testing*.

3. presuppose that students will need calculators while taking the test. The test is developed so that, for a majority of students, some portion of the items *require* calculator use in order to be solved successfully. An appropriate term for this approach is *calculator-based testing*. (p. 157)

Note that "calculator-passive testing" requires no changes in the mathematics tests that are presently administered to measure student achievement or aptitude. However, letting students use a calculator very likely will change what is being tested. "Calculator-neutral" tests are tests that permit but do not require those taking them to use calculators, and do not include any items on which calculator use will benefit test takers. "Calculator-based testing" assumes students have available and can use calculators effectively as tools.

In an earlier context Harvey defined calculator-based mathematics tests and calculator-active test items as follows:

A *calculator-active test item* is an item that (a) contains data that can be usefully explored and manipulated using a calculator and (b) has been designed to require active calculator use.

A *calculator-based mathematics test* is one that (a) tests mathematics achievement, (b) has some calculator-active test items on it, and (c) has no items on it that could have been, but are not, calculator-active *except for* items that are better solved using non-calculator based techniques. (Harvey, 1989, p. 78)

Note that my use of the term, "technologically-active tasks," is based on this definition. Harvey notes at one point: "It seems quite likely that many calculator-based tests have been developed; however, not many of these tests have been widely circulated or discussed" (1992, p. 158). The only tests he found were from a single dissertation study, the tests being developed by the MAA Calculator-Based Placement Test Program Project, and the chapter tests developed by the Ohio State Calculator and Computer Precalculus Curriculum (C^2PC) Project (Demana & Waits, 1989). The tests and test items that have been produced by the MAA Calculator-Based Placement Testing Project and the Ohio State C^2PC Project demonstrate that valid, reliable calculator-based tests and calculator-active items can be generated that satisfy the definitions of the terms that were given earlier in this paper. However, at present, there is a paucity of published calculator-based tests and calculator-active items. Furthermore, assessment tasks where students are expected to use other technologies are non-existent in the literature. Given the variety of curriculum projects being developed where students are to use some technological tools other than paper and pencil, it is surprising that so few "technologically-active" assessment tasks are available.

Examples of Calculator Items from NAEP

To conclude this paper, I have chosen to present four items from the calculator-based subtest of the 12th-grade and one item from the 4th-grade NAEP tests administered in 1990. For each item, note that the student was expected to answer the question--"Did you use the calculator on this question?" Also, the p-value for the item is given. For 12th-grade students, the calculator would seem to be of little use on items 3 or 19. Item 9 is better, but why is an answer "to the nearest hundredth" necessary? Item 15 is clearly a calculator-active item, although a thoughtful student could guess the correct answer without knowing how to calculate the answer. This would have been better in an open format. The 4th-grade calculator item is nonsense. Finally, I have included one 12th-grade item in a subtest where calculators were not allowed--an item in which many students might have found a calculator useful. I can only conclude that looking to NAEP for good calculator-active items is not warranted.

3. The figure above shows the display on a scientific calculator. The value of the displayed number is between which of the following pairs of numbers?

  Ⓐ 0.04 and 0.05

  Ⓑ 0.4 and 0.5

  Ⓒ 4.0 and 5.0

  Ⓓ 40.0 and 50.0

  Ⓔ 400.0 and 500.0

Did you use the calculator on this question?

○ Yes   ○ No

NAEP 12th-Grade 1990
Item 3 (p = 29.7)

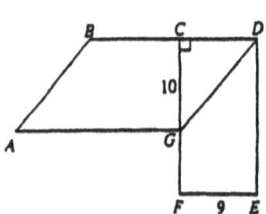

9. In the figure above, $ABDG$ is a parallelogram and $CDEF$ is a rectangle. If $EF = 9$ and $CG = 10$, what is AB to the nearest hundredth?

Answer: _____

Did you use the calculator on this question?

○ Yes   ○ No

NAEP 12th-Grade 1990
Item 9 (p = 20.8)

15. A savings account earns 1 percent interest per month on the sum of the initial amount deposited plus any accumulated interest. If a savings account is opened with an initial deposit of $1,000 and no other deposits or withdrawals are made, what will be the amount in this account at the end of 6 months?

(A) $1,060.00

(B) $1,061.52

(C) $1,072.14

(D) $1,600.00

(E) $6,000.00

Did you use the calculator on this question?

○ Yes   ○ No

NAEP 12th-Grade 1990
Item 15 (p = 14.9)

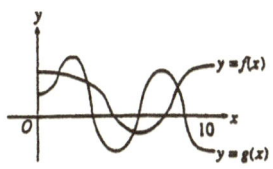

19. The graphs of $y = f(x)$ and $y = g(x)$ for $0 \leq x \leq 10$ are shown in the figure above. For how many values of x is the product $f(x)g(x) = 0$ for $0 \leq x \leq 10$?

(A) Two

(B) Four

(C) Five

(D) Six

(E) Seven

Did you use the calculator on this question?

○ Yes   ○ No

NAEP 12th-Grade 1990
Item 19 (p = 40.6)

5. The third grade collected more than 850 bottle caps for an art project. The fourth grade collected more than 500 bottle caps. Using her calculator, Maria found the exact total of all the bottle caps collected by both grades. Which calculator could be hers?

Ⓐ 350.

Ⓑ 850.

Ⓒ 1284.

Ⓓ 1376.

Did you use the calculator on this question?

○ Yes   ○ No

NAEP 4th-Grade 1990
Item 5 (p =.60)

TELEPHONE CALLING RATES

From Allenville To	Day Rate 8 AM–5 PM Mon–Fri		Evening Rate 5 PM–11 PM Mon–Fri 8 AM–11 PM Sat–Sun		Night Rate 11 PM–8 AM ALL DAYS	
	First Minute	Each Additional Minute	First Minute	Each Additional Minute	First Minute	Each Additional Minute
Burneyford	$.09	$.03	$.07	$.02	$.05	$.02
Camptown	$.28	$.09	$.22	$.07	$.17	$.05
Dorning	$.37	$.11	$.30	$.09	$.22	$.07
Edgeton	$.42	$.12	$.34	$.10	$.25	$.07

8. The table above provides information about the cost of placing phone calls between certain cities at different times during the day. How much more would it cost to place a 10-minute call from Allenville to Edgeton at 3 pm on Friday than at 3 pm on Saturday?

Answer: _____

NAEP 12th-Grade 1990
Item 8 (p = 31.0)

Summary

My intent in this paper is to raise some questions about assessment, the role of technological tools, and "calculator-active" assessment tasks. The questions were addressed, but not answered; the role of tools discussed, but not clarified; and the notion of "technologically-active" tasks defined, but a wealth of examples not found. Thus, there is clearly a lot of work to be done.

References

Archbald, D., & Newmann, F. (1988). *Beyond standardized testing: Assessing authentic academic achievement in the secondary school.* Reston, VA: National Association of Secondary School Principals.

Bodin, A. (1991). *Replication et complements, 1991, des evaluations Sixieme et Cinquieme.* Irem: Universite de Besancon.

Campbell, P. F., & Fey, J. T. (1988). New goals for school mathematics. In R. S. Brandt (Ed.), *Content of the curriculum: 1988 ASCD yearbook* (pp. 53-74). Alexandria, VA: Association for Supervision and Curriculum Development.

Char, C., Hawkins, J., Wootten, J., Sheingold, K., & Roberts, T. (1983). *"Voyage of the Mimi": Classroom case studies of software, video, and print materials.* New York: Bank Street College of Education, Center for Children and Technology.

Corno, L., & Snow, R. E. (1986). Adapting teaching to individual differences among learners. In M. C. Wittrock (Ed.), *Handbook of research on teaching* (3rd ed., pp. 605-629). New York: Macmillan.

Demana, F., & Waits, B. K. (1989). *Precalculus mathematics: A graphing approach.* Reading, MA: Addison-Wesley.

Donovan, B. F., & Romberg, T. A. (1987). Knowledge structures and assessment of mathematical understanding. In T. A. Romberg & D. M. Stewart (Eds.), *The monitoring of school mathematics: Background papers. Vol. 2: Implications from psychology; Outcomes of instruction* (pp. 225-242). Madison: Wisconsin Center for Education Research, School of Education, University of Wisconsin-Madison.

Edwards, L. D. (1991). Children's learning in a computer microworld for transformation geometry. *Journal for Research in Mathematics Education, 22,* 122-137.

Frederiksen, N. (1984). The real test bias: Influences of testing on teaching and learning. *American Psychologist, 39,* 193-202.

Goodson, I. (1988). *The making of curriculum.* London: Falmer Press.

Harvey, J. G. (1989). What about calculator-based placement tests? *The AMATYC Review, 11*(1, Part 2), 77-81.

Harvey, J. G. (1992). Mathematics testing with calculators: Ransoming the hostages. In T. A. Romberg (Ed.), *Mathematics assessment and evaluation* (pp. 139-168). Albany, NY: State University of New York Press.

Jones, L. V. (1988). School achievement trends in mathematics and science, and what can be done to improve them. In E. Z. Rothkopf (Ed.), *Review of research in Education, 15, 1988-89* (pp. 307-341). Washington, DC: American Educational Research Association.

Kaput, J. J. (1989). Linking representations in the symbol systems of algebra. In S. Wagner & C. Kieran (Eds.), *Research issues in the learning and teaching of algebra* (pp. 167-194). Reston, VA: National Council of Teachers of Mathematics and Lawrence Erlbaum Associates.

Kepner, H. S. (1989). *Graphics calculators: A step in the democratization of mathematics - beyond arithmetic.* Unpublished manuscript, University of Wisconsin-Milwaukee.

Kirkland, M. C. (1971). The effects of tests on students and schools. *Review of Educational Research, 41*, 303-350.

Mislevy, R. (in press). Foundations of a new test theory. In N. Frederiksen, R. Mislevy, & I. Bejar (Eds.), *Test theory for a new generation of tests.* Hillsdale, NJ: Erlbaum.

National Council of Teachers of Mathematics. (1989). *Curriculum and evaluation standards for school mathematics.* Reston, VA: Author.

National Council of Teachers of Mathematics. (1991). *Professional standards for teaching mathematics.* Reston, VA: Author.

National Council on Education Standards and Testing. (1992). *Raising standards for American education.* Washington, DC: U.S. Government Printing Office, Author.

National Research Council. (1989). *Everybody counts: A report to the nation on the future of mathematics education.* Washington, DC: National Academy Press.

Oakes, J. (1986). *Educational indicators: A guide for policy makers.* Santa Monica, CA: Center for Policy Research in Education, The RAND Corporation.

Pea, R. D. (1987). Cognitive technologies for mathematics instruction. In A. H. Schoenfeld (Ed.), *Cognitive science and mathematics education* (pp. 89-122). Hillsdale, NJ: Lawrence Erlbaum Associates.

Resnick, L. (1987). *Education and learning to think.* Washington, DC: National Academy Press.

Romberg, T. A. (1988). Evaluation: A coat of many colors. *Evaluation and assessment in mathematics education.* UNESCO Document Series No. 32.

Romberg, T. A. (1992). Problematic features of the school mathematics curriculum. In P. W. Jackson (Ed.), *Handbook of research on curriculum* (pp. 749-788). New York: Macmillan.

Romberg, T. A., & Wilson, L. (in press). Issues related to the development of an authentic examination system for school mathematics. In T. A. Romberg (Ed.), *Assessment in school mathematics.* Madison: National Center for Research in Mathematical Sciences Education.

Schwartz, J. L., & Yerushalmy, M. (1985). *The Geometric Supposer* s [computer software and teachers' guide]. Pleasantville, NY: Sunburst Communications.

Schwartz, J. L., Yerushalmy, M., & Education Development Center (1988). *Visualizing algebra: The function analyzer* (Computer software and teachers' guide). Pleasantville, NY: Sunburst Communications.

Stake, R. (in press). The validity of standardized testing for measuring mathematics achievement. In T. A. Romberg (Ed.), *Assessment in school mathematics.* Madison: National Center for Research in mathematical Sciences Education.

U.S. Department of Education. (1991). *America 2000: An education strategy.* Washington, DC: Author.

Webb, N. L. (1987). Another look at assessment: A reaction to chapters 17-20. In T. A. Romberg & D. M. Stewart (Eds.), *The monitoring of school mathematics: Background papers; Vol. 2: Implications from psychology; Outcomes of instruction* (pp. 243-260). Madison: Wisconsin Center for Education Research, School of Education, University of Wisconsin-Madison.

Webb, N. L., & Romberg, T. A. (1988, April). *Implications of the NCTM standards for mathematics assessment.* Paper presented at the annual meeting of the American Educational Research Association, New Orleans.

Zarinnia, E. A., & Romberg, T. A. (1987, March). A new world view and its impact on school mathematics. In T. A. Romberg & D. M. Stewart (Eds.), *The monitoring of school mathematics: Background papers. Vol. 1: The monitoring project and mathematics curriculum* (pp. 21-62). Madison: Wisconsin Center for Education Research.

Calculator Inservice for Mathematics Teachers

George W. Bright
University of North Carolina at Greensboro

The rapid development of calculator technology has created a gap between the way that school mathematics is typically taught without calculators and the way that school mathematics might be taught in the presence of calculators. In order to bridge this gap teachers need to investigate uses of calculators first in actually "doing" mathematics and then in communicating mathematical ideas. That is, teachers need inservice experiences not only in learning mathematics with the aid of calculators but also in ways of teaching mathematics when calculators are available for students to use.

Proper use of calculators would exploit their unique capabilities to teach both old and new mathematics in new ways. Calculators should not be used merely to deliver the same mathematics instruction through an alternate media. Historically, teachers and researchers have typically sought answers to the question of whether technology improves learning (e.g., Salomon & Gardner, 1986). However, just asking if calculators can help teachers do things better is almost certainly not the best question to raise. Rather, we need to know how to use the unique capabilities of calculators to improve the learning of content. That such capabilities are in fact unique mitigates against the possibility even of comparing the new ways with "standard" instructional techniques; for example, by carrying out standard experimental/control group experimental designs. Teaching techniques involving calculators are likely to be non-comparable to other kinds of instruction.

At the secondary level, Fey (1984) called attention both to new mathematical ideas that can be taught when computers are used and to new instructional techniques that can be developed using computing technology. The whole body of questions, concerns, and ideas discussed there almost constitutes a reconceptualization of secondary mathematics teaching. Does there need to be a similar reconceptualization for mathematics based on existing and emerging calculator technologies? If so, does there need to be a different reconceptualization for each type of calculator or for each significant functionality of calculators? Would these reconceptualizations interact with grade level? Answers to these questions are fundamental for developing our understanding of the impact of calculators on instruction, and our answers will have significant and direct impact on the design and implementation of inservice programs for mathematics teachers.

Calculator Inservice in Context

Inservice on the use of calculators in teaching mathematics is part of the larger picture of inservice on technology. Although there are undoubtedly commonalities of inservice approaches across all technologies (e.g., providing hands-on time working with a technology), there appear to be some concerns unique to programs built around calculators (e.g., concerns about adverse effects of calculator use on students' learning of mathematics). Calculator inservice also sits within the broader picture of inservice in general. Much effort has been directed at understanding ways to improve the impact of inservice programs.

One model, the Concerns Based Adoption Model (CBAM) developed at the R&D Center at the University of Texas in the 1970s (e.g., Hall, 1976; Hall & George, 1979; Hall, Wallace, & Dossett, 1973), is a means of tracking the concerns of teachers as they try to implement an innovation in instruction. Much of the research on CBAM has dealt with implementation of specific programs (e.g., locally designed reading program), new techniques for teaching standard content (e.g., hands-on science), or general techniques for improvement of teaching (e.g., peer coaching) rather than with the reorganization of curriculum and pedagogy based on a new technology. New research using CBAM (Friel & Joyner, 1991) is beginning to investigate its appropriateness in a context of substantive change; this research takes a general perspective rather than an orientation of implementation of a specific program. The results of this effort may help us understand how such general interventions, like that created by calculators, might affect instruction.

Another model for understanding teaching is the novice/expert dimension described by Berliner (1988). Teachers seem to pass through several stages of development as they learn to cope with real classroom situations: novice, advanced beginner, competent teacher, proficient teacher, expert teacher. This perspective can be applied to a teacher's use of technology (e.g., Lamphere, Usnick, & Bright, 1989) and, hence, to the use of calculators.

Finally, there is some evidence that the involvement of teachers in the development of instructional materials based around an innovation may be important in order for that innovation to be sustained over a long term in classroom instruction. Participation in curriculum development gives teachers a sense of (a) helping to shape instruction, (b) responsibility for the innovation, and (c) commitment to making the innovation succeed (Morin, 1986; Young, 1985). Teachers' involvement in curriculum development appears to be a more critical factor than the quality of the curriculum developed in sustaining an innovation (Berman & McLaughlin, 1976). It is unclear how extensive a teacher's involvement need be, but we have some evidence (e.g., Prokosch, Bright, & Freiberg, 1991) that involvement in adapting materials may be adequate to change what goes on in the classroom when calculators are incorporated in instruction. The existing research base does not extend over a long enough term, however, to be sure that calculator-based instruction will continue once support of teachers is withdrawn.

Areas of Concern for Calculator Inservice

There are several different areas of concern that any calculator inservice program must address. These include some standard concerns, which might be conceptualized as low-level concerns: (a) calculator skill, (b) understanding of relevant mathematics, and (c) knowledge of pedagogy specific to the incorporation of calculators in instruction. But there are also more sophisticated concerns, which might be conceptualized as high-level concerns: (a) interaction between calculator use and testing, and perhaps more importantly (b) teachers' beliefs about the role of calculators in learning mathematics.

Low-Level Concerns

Calculator skill. Although it is clear that any inservice on calculator use needs to include instruction on the use of the particular keys for the calculators being used, very little seems to have been written about this concern. Many people (e.g., Bright, Usnick, & Lamphere, 1989; Osborne, 1991) give attention to introducing keystrokes to teachers, but the over-riding assumption seems to be that knowledge of key strokes will develop as a calculator is used. This is similar to the approach apparently taken by the University of Chicago School Mathematics Program with students (Hirschhorn & Senk, 1992).

It would seem to be important, however, for inservice leaders to monitor whether lack of "keyboarding" skill is getting in the way of the learning of either content or pedagogy. In particular, my experiences with middle school teachers suggest that at least initially, they are overwhelmed by all of the functionality of scientific or graphing calculators, even though most of that functionality is not even demonstrated during an introductory workshop. They seem to believe that they are supposed to know all of the functionality immediately, and they are quite uncomfortable when they don't know what particular keys do.

Similarly, teachers need to be warned that lack of keyboarding skill might be an area they will need to address with their students. However, teachers need to be clear about whether the lack of skill is important for the mathematics being taught or whether students react like teachers and become overwhelmed because of what they don't know about any particular calculator.

Understanding of mathematics. It seems clear that the type of mathematics that needs to be (or ought to be) taught when calculators (especially sophisticated calculators) are present will be different than the historical curriculum of school mathematics. Teachers need to recognize that the main goal of mathematics instruction needs to shift away from learning about procedures to learning about conceptual ideas useful for problem solving. This is a radical shift. Resnick (1987) points to a similar shift across all content; namely, expecting *all* students to learn to think. It may be of some comfort to know that mathematics teachers are not alone in dealing with greater expectations.

As long as teachers continue to perceive that their goal is to get students to understand content (e.g., numeration, algebra, geometry) rather than applications of that content (e.g., problem solving), they are going to find integration of calculators difficult, since they will likely view calculator use as another "add on" to the curriculum. Although calculators can be useful for learning content, they may be most useful for solving problems. Calculators give users power over the world. The ground swell for significant change in mathematics instruction toward reaching this problem solving focus will come only if teachers agree, and buy into the fact, that students really will have to know how to use calculators to survive in the world. Calculators are essential for understanding the relevance of mathematics in today's world.

It may be important here to note that teachers neither easily nor quickly accept the notion that calculators are essential to mathematics learning and mathematics teaching. Teachers typically take several years to internalize this view. I can point to two major projects with middle school teachers at the University of Houston on this point; both projects extended over several years, and significant change in classroom instruction did not happen until well into each project. Even at the university level, numerous anecdotes about university faculty point to a similar period of adjustment for these mathematically-sophisticated people. It takes all teachers a long time to learn what mathematics is appropriate when calculators are available for students to use.

Confounding this is the fact that many teachers appear afraid that if students have calculators and explore mathematics themselves, then a teacher's deficiencies in understanding of content may be revealed. Some teachers appear afraid of losing their traditional role in the process of instruction; namely, the role as primary authority about the content. Having calculators available for learning mathematics seems risky to some teachers.

In any event, it seems quite important for calculator inservice programs to help teachers learn some (new-to-them) mathematics in the presence of calculators. Such opportunities allow teachers to experience both the frustration and the exhilaration of understanding the output that appears in a calculator's display. Indeed, calculator inservice has in some cases been an excuse for increasing the mathematical knowledge of teachers. Learning of content can fairly easily be "hidden" in activities whose surface goal seems to be learning how to use calculators to solve problems. I prefer to use inservice on calculators (and indeed, on other technologies, too) as an opening or motivation for a whole variety of other goals; for example, learning mathematics content and changing the pedagogy used in teaching mathematics. It seems important for teachers to identify for themselves the ways that calculator use needs to affect what they know and what they do.

An example of the ways that the view of mathematics changes is that graphing calculators create the need for understanding of a complete graph; namely, understanding of the global properties of a function. Changing the scale can dramatically influence the picture of the function that is actually visible. Teachers often aren't familiar with the notion of complete graph, and the scale typically used in paper-and-pencil graphing typically doesn't require this knowledge. When graphing calculators are used in instruction, teachers have to rethink the entire notion of graphical representation. Teachers find themselves asking questions which they haven't asked before about the mathematics of functions, and they often feel unsure of what they know and do not know.

Many areas of mathematics will be handled similarly as teachers learn to use different types of calculators effectively. Calculators seem to cause changes in the ways that people look at mathematics and the ways they work problems.

Knowledge of pedagogy. Probably the most significant change in mathematics pedagogy when calculators are used is that teachers and their students ask different questions. Calculators allow even lower-level students the chance to talk in very sophisticated ways about mathematics concepts. But teachers need to figure out new ways to ask students to deal with underlying concepts. Inservice activities can assist this process, both by modeling some questioning strategies for teachers and by allowing teachers to muck around for themselves until good ways to ask questions emerge. Sometimes (and more often these days, it seems, as the technology base continues to shift under our feet) teachers just have to stumble on to the "right" question. We just don't know what the "right" questions are ahead of time.

When calculators are used, teachers often have to shift from worrying about whether the student has the right answer to worrying about whether the teacher has asked the right question. Currently, teaching seems focused on the products of students. The real test of effective teaching with calculators, however, may be whether teachers have asked appropriate questions to help students generate deep understanding of the mathematics. It is important to help students make connections between what they see in the calculator display and the symbolic (often algebraic) representation of the mathematics.

Teachers' roles (and, by extension, workshop leaders' roles and university faculty roles) must also change significantly when mathematics is taught with calculators. In particular, calculators empower students to explore. It is not unusual for students to become the arbiters of when a teacher's help is needed. That is, students can say, "I'll let you know when I need help. Until then, just stay out of my way." Students now have the chance to experiment and to learn, sometimes in spite of the teacher. In workshops, teachers should take on students' roles and become responsible for their own explorations.

It is important also to acknowledge the role of support materials in inservice that focuses on pedagogy. These materials are critical models for helping teachers begin to think about the range of possibilities for changing pedagogy. Although support materials for computer software are getting better, support materials for calculators are often quite pedantic. Classroom teachers need good materials to use in instruction, that is, materials that teach important mathematics clearly. These materials give teachers confidence to actually use calculators the first few times with students. We can't say to teachers, "Go back and use the same textbook, only now with calculators." We have to change the print support materials so that the unique capabilities of the calculators are used meaningfully.

The goal of the middle school project that I directed in the late 1980s was to change pedagogy by helping teachers learn to use calculators and computers effectively. Teachers were also expected to produce teaching materials built around technology and to become models within their schools of the use of technology in teaching mathematics. As alluded to earlier, the requirement to produce or adapt teaching materials may have been critical for the success of the project, though at the time, I did not recognize how important it was. The project had several components: (a) academic course work in summer 1988, (b) additional course work, project meetings, and classroom visitations to support use of technology in teaching mathematics in the 1988-89 academic year, (c) more courses and a writing workshop in summer 1989, and (d) continuing writing/revising workshops in fall 1989. The courses were designed to provide the information the teachers needed to expand their perspectives of both mathematics and technology and to integrate technology into the teaching of mathematics. The work during the 1988-89 academic year was designed to monitor the extent of actual integration of technology, with encouragement given as needed. The writing workshops in summer and fall 1989 were designed to provide a synthesis for the teachers and to provide an opportunity to evaluate the success of the integration activities. Three project evaluation techniques revealed important information about pedagogical knowledge: (a) pre/post interviews with participants, (b) surveys completed at the end of the project, and (c) formal observation of students and teachers while technology was used in mathematics instruction.

The pre-interviews indicated that virtually all of the participants had no experience using technology to teach mathematics. The limits of their experience with technology was generally limited to four-function calculators and computers in drill and practice mode. Many felt that technology was not likely to be as effective in helping students learn mathematics as other teaching techniques. Several participants expressed the concern that students would not learn the "basics" if calculators were used too often, and their views of appropriate uses of calculators was typically "to check homework."

The post-interviews showed a very different picture. All teachers were confident of their ability to use some technologies in teaching mathematics, though only a few participants were comfortable with a wide range of technologies. They generally acknowledged that technology was useful in developing conceptual understanding and that their role as teachers was to guide this conceptual development.

The surveys revealed several interesting changes related to pedagogical knowledge. Three of the "stages of concern" items (i.e., computer education research, calculators, and computers) showed "growth" of more than three stages (Hall & George, 1979). These related more to pedagogical concerns (i.e., understanding of both how to use technology and what the effects of those uses are on student learning) than to mathematics content concerns. The two items that showed the greatest change in perceived pedagogical expertise were (a) using calculators in teaching mathematics and (b) designing curriculum materials. These two items seem highly relevant to a "regular" mathematics teacher who has limited access to technologies for teaching.

Formal classroom observations were conducted between March and May, 1990, after the teachers had spent five semesters studying ways of using technology in teaching middle school mathematics. Thirteen teachers were each observed once while using technology to teach mathematics; 6 of the teachers used four-function calculators only, 1 used scientific calculators only, 1 used four-function calculators and computers together, 1 used fraction calculators and computers together, and 4 used computers only. Although computer lessons were more effective than calculator lessons at keeping students on-task, it is more interesting to compare the technology lessons with "traditional," non-technology lessons.

For the 13 classes, students were on-task an average of 97% of the time, independent of the type of technology being used. Instruction activities were coded as lecture/demonstration (36.5%), questioning (36.5%), seat work (8.6%), and other (18.3%); knowledge level was coded as conceptual (89.5%) and procedural (10.5%). These data suggest a classroom environment that does not match the stereotypical mathematics lesson in which a teacher asks questions on homework, briefly presents new material (which is often times a procedure), and then lets students work on assigned exercises to practice the procedure. The project teachers seemed to be trying to engage students (i.e., questioning) in order to help them understand the mathematics (i.e., conceptual level). The teachers appeared to give minimal importance to students working on their own (i.e., seat work) to master procedures. It seems that technology was a vehicle for significant reconceptualizing of appropriate pedagogy. (For more information, see Prokosch, Bright, & Freiberg, 1991.)

High-Level Concerns

Testing. One difficult issue for teachers to deal with is testing. We are so used to writing particular kinds of test questions (namely those with exactly one right answer or with no need for exploration of ideas) that it is difficult to break our mind set. It seems important for us to think about test questions that allow students to do things they couldn't do (or haven't done) before. If we're going to use calculators, we ought to test in ways that allow students the chance to show what they have learned.

Consequently it is critical that inservice programs allow teachers chances to talk about assessment. These discussions will naturally raise issues of what is important in the curriculum. Indeed, the two notions of what to test and what to teach cannot be separated for very long. It also seems critical for discussion to occur in several parts across several sessions, for as teachers learn more about the use of calculators, their views of assessment and curriculum will develop.

Teachers also frequently complain that they are held accountable by parents, school boards, state departments of instruction, and legislatures for performance on standardized, non-calculator tests. Yet, many of these same groups often expect that instruction will include attention to the use of calculators and other technologies. The discrepancy between those seemingly mixed messages creates quite legitimate concerns. Fortunately, more and more states are recognizing this discrepancy and are trying to deal with it through the creation of new tests. In North Carolina, for example, there are end-of-grade tests for grades 3-8 and end-of-course tests for many of the high school subjects. Beginning in 1993, about 85% of the items on tests in grades 3-8 will be written with the assumption that calculators are available during testing. The particular calculators to be used are four-function for grades 3-5 and fraction calculators for grades 6-8. The Algebra I end-of-course test, which is expected to be implemented in 1994, will assume the availability of graphing calculators. This is, as you might expect, causing a great upsurge of interest among teachers on ways to teach with calculators.

Beliefs. Perhaps the most important filter for teachers about the integration of calculators into instruction is their beliefs about when calculators are or are not appropriate for students to use. Teachers who believe that the primary use is for checking work can be expected to use calculators differently than teachers who believe that students should use calculators during initial learning of a concept. Changing teachers' beliefs is obviously a long process, with no guarantee that any attempt to do so will be successful. My suspicion is that a necessary part of this process is having teachers learn mathematics with the assistance of calculators. But that is clearly not a sufficient condition.

During the 1991 NCTM/SIG research presession, John Harvey recounted two anecdotes of his attempts to get both undergraduate and graduate mathematics students to decide on appropriate ways to use technology in teaching.

> Two or three semesters ago I taught the [computer education] curriculum course. [The students were] secondary teachers [who had completed] a semester of computer programming. They were nearing the end of their teacher education, and I decided that we should discuss the future curriculum. And so I found some pieces of software including Master Grapher and its accompanying text materials and I divided them into groups and

gave each one of them a piece of software and told them their responsibility was twofold. One was to learn how to use the software, and the second was to come up with some very specific suggestions of where that particular piece of software could be used in teaching mathematics in [grades] seven through twelve. I will not say that my experiment was a failure, but it was not very successful. It became apparent that the amount of time needed for mathematically wise students to learn the technologies is long. In other words, it's not spontaneous. ... As a matter of fact, a group of three young men ... came back and they did some very fancy things with Master Grapher. They had had the text materials for a three or four week period, and they concluded that this piece of software should not be used in secondary schools with students because it was too hard to use... It was too complicated and too hard to use. And it speaks to beginning to teach. They've got to do mathematics the way in which we're expecting them to teach mathematics.

I find it very interesting at Wisconsin that the graduate students in mathematics many of them minor in statistics or in computer science. So they are very technologically oriented. ... But when you talk to them about how they're going to teach their undergraduate courses, when they graduate; or as they're presently teaching pre-calculus and ask them if they'll use technology, they will say "Definitely not, because if you use technologies kids won't learn mathematics." (Harvey, 1991)

The reluctance of these prospective teachers may have been significantly colored by the lack of use of technology in their own mathematics courses, but if we wait to reform school mathematics until all teachers have used calculators in their own mathematics learning, we'll never reach the goal of integrating calculators into instruction. We must find other ways of helping existing teachers see the importance of using calculators in their instruction.

CBAM research suggests that it will take most teachers a long time to work their way through the various types of concerns that will ultimately result in concern about students' learning. Inservice that addresses such concerns too early, that is, before teachers are ready to deal with them, is not likely to be effective. Acknowledging the need for match between teachers' concerns (i.e., their underlying beliefs structure) and the focus of inservice programs would seem to be the first step toward successfully moving teachers toward the goal of integrating calculators into instruction.

Effects of Students on Teachers and Teaching

Integrating technology into instruction is unlike most other instructional innovations along one key dimension. Technology gives students control of their learning to an extent unlike any other innovation. Calculators are only slightly less dramatic than computers in giving students the power to explore on their own. In the primary grades, it doesn't take children long to get a calculator to display a decimal or a negative number. Once this happens, they are often insistent that the teacher help them understand what this new symbol means. In high school, students want to know what the graph of $y=x^{10}$ or $y=x^{100}$ looks like, and then they won't want to restrict attention "merely" to quadratics. That is, the content that teachers have to deal with gets sophisticated very quickly, simply because calculators let students display information about concepts that the traditional curriculum has never included. Some teachers find this upset of the traditional sequence quite discomforting, while others embrace it with open arms - and open minds. My personal bias is that we want all teachers not only to embrace students' willingness to explore but in fact to encourage it. I don't know, however, how to structure inservice programs to help all teachers reach this level of flexibility quickly.

At the preservice level, there is the additional difficulty that cooperating teachers who are serving as mentors for student teachers often are not modeling calculator use and are unwilling to let student teachers experiment with calculator use, either because the cooperating teacher does not believe that calculators are appropriate or because the cooperating teacher is simply unsure of the effects of such explorations. Perhaps one of the best things we could do for preservice teachers is to convince them that the first things they ask for in their first job are calculators and computers. We need to be sure that they understand that mathematics simply can't be done without these tools.

On the other side of the coin, there are some students who believe that using calculators is cheating; and there are probably lots of parents with similar beliefs. Students may even feel guilty about doing mathematics with a calculator because they think it will be too easy. Students sometimes take pride in being able to carry out complex procedures, even (or maybe especially) if they do not understand the mathematics behind those procedures. Part of any successful inservice (or preservice) program will include preparing teachers to deal with students' and parents' beliefs about when calculator use is appropriate. And just as it will take time for teachers to adjust their value systems, it will take students and parents time to adjust theirs. At least in my experience, however, one of the most effective arguments is to focus on the increase in higher level thinking that students are able to engage in when calculators are available.

Networking

Research on inservice suggests that networking among teachers is a critical factor in successful implementation of an innovation. However, I would argue that it is even more critical for successful integration of calculators into instruction. When teachers decide to use calculators, they have to learn calculator skills, pedagogy skills, and often times new mathematics. It is difficult for one person to accept all of this responsibility personally. It is important for teachers to have a support system in place to help deal with all of their concerns.

There are many ways to build camaraderie. Some programs use social activities (e.g., picnics, sports, beer drinking) while other programs use more academic pursuits (e.g., sharing of materials developed for student use). The most critical goal is that teachers identify people that they feel comfortable contacting when problems arise. The contact can be in person, by telephone, or through electronic means (e.g., e-mail). But having a buddy who has "survived" the same implementation struggles seems to be important.

The particular techniques used to establish teacher networks may depend to some extent on the nature of the group of teachers participating. Evaluation of one of my own programs (Bright, 1991) suggests that teachers who have an established peer network in their buildings seek different kinds of support from workshops and from colleagues than teachers who are more isolated in their teaching environments. Teachers with existing support want information that they can share with others in the existing network, while teachers without existing support look for models that they can adopt quickly and independently in their own classrooms.

Implementation of Calculator Instruction

There are several approaches that might be taken to helping teachers implement calculator-based instruction. One approach is to deal with teachers at the local level (e.g., Super, 1992). The most obvious advantage is that the inservice can be tailored to local needs, while a obvious disadvantage is that many people end up "reinventing the wheel" as they invest time in planning inservice efforts for relatively few people.

A second approach that has advantages of scale is an inservice for a much larger group of teacher, such as all the teachers in a state (e.g., Bright, Lamphere, & Usnick, 1992). Investing somewhat more resources in the planning phase of this type of program might result in higher quality materials, which would address common themes but would not respond as well to unique local needs. The effectiveness of such large efforts seems to depend on the quality of the materials and on the organizational structure established for managing the effort. Teachers have quite effective "grapevines" for sharing information. Attempts to implement poor quality inservice would soon meet with considerable disinterest from teachers as they shared their reactions, while attempts to implement high quality inservice can be expected to meet with large numbers of teachers willing to sign up for the program. The evaluation of the Texas Mathematics Staff Development Program (Randall & Bennett, 1991) supports this view, though more data from more large-scale programs would be helpful. No matter how good the inservice might be, however, if the institutional structure to manage the delivery process is inadequate, teachers will become frustrated with bureaucratic hassles and may refuse to participate. Advertising of inservice sessions, easy registration procedures, and easily accessible sessions all seem important to teachers.

At both local and statewide levels, some calculator inservice programs have tied delivery of classroom sets of calculators to participation by the teacher in the inservice program. This seems to be an effective way to entice teacher participation in inservice. Indeed, teacher inservice leaders frequently share anecdotes that suggest that teachers are very interested in getting materials and resources for their classrooms, even to the point that handing out materials may be as effective at attracting teachers to an inservice program as providing stipends for participation. More careful documentation of the effects of requiring inservice participation as part of making calculators available for students to use would seem to be needed, however. I don't think that we want to leave the impression that a particular program is the ultimate inservice program on the use of calculators. But we do want to be sure that teachers understand that putting calculators into their classrooms will change not only the mathematics that is taught put also the pedagogy that is used. Teachers have to understand that they need to make a commitment to learning about the effects that these technological tools will have.

For both local and larger-scale implementation efforts, it is necessary to develop a base of support from administrators and parents. Without that support teachers can be left hanging with interest but with little assistance. As part of the statewide effort in Texas (with funding from the Eisenhower Education Act National Programs) a training module for administrators and school board members was developed (Dockweiler & Schielack, 1989). Workshop leaders who have subsequently been trained to deliver the Texas Staff Development Modules have also been trained to deliver the administrator/school board module. Their feedback has been that it is effective at raising consciousness among administrators and that it helps to generate support for purchase of the technology that teachers want to use. I would urge other calculator inservice programs to consider the necessity for developing similar administrative support for acquisition of the calculators teachers need to teach mathematics with calculators.

Conclusions

Integrating calculators into mathematics instruction is an important and difficult goal to accomplish. Inservice programs must extend over a long period of time and provide significant support for teachers throughout that time. The main goal of that support is to give participants enough confidence to begin experimentation with the use of calculators in their teaching. Once teachers try using calculators and see the resulting enthusiasm of the students, they are more likely to continue to use calculators. Teachers are very concerned about motivating their students to learn, and calculators and other technologies are very powerful motivation. The initial hurdle, however, is to get teachers to use calculators the first few times. Then teachers may become more concerned about actual learning outcomes. Fortunately, research evidence is fairly clear that student learning will almost universally improve (or at least not decline). Administrators who come in to observe teaching may be more impressed with the improvements in time-on-task that are typical when technology is integrated in instruction than with the actual use of the technology. It may be more problematic when these evaluators see a relatively higher level of student-initiated activity and a relatively lower level of teacher lecture/demonstration. Inservice programs need to help teachers learn how to reeducate the evaluators about what kinds of behaviors are appropriate in a classroom where calculators are used extensively.

Inservice programs need to have continued and increasing use of calculators as their long term goal. Inservice can also help teachers become more sophisticated in their uses of calculators, but only if we are willing to work *with* teachers in their search for more effective mathematics instruction. Effective calculator inservice requires that we commit our energy and resources to the long-term effort needed to effect these changes.

References

Berliner, D. C. (1988, February). *The development of expertise in pedagogy.* Paper presented at the annual meeting of the American Association of Colleges for Teacher Education, New Orleans, LA.

Berman, P., & McLaughlin, J. (1976). Implementation of educational innovation. *Educational Forum, 60*, 345-370.

Bright, G. W. (1991, April). *Technology inservice for middle school mathematics teachers.* Paper presented at the annual meeting of the American Educational Research Association, Chicago, IL.

Bright, G. W., Lamphere, P., & Usnick, V. E. (1992). Statewide in-service programs on calculators in mathematics teaching. In J. T. Fey & C. R. Hirsch (Eds.), *Calculators in mathematics education: 1992 yearbook* (pp. 217-225). Reston, VA: National Council of Teachers of Mathematics.

Bright, G. W., Usnick, V. E., & Lamphere, P. (1989). *Calculators for grades 6-10.* Austin, TX: Texas Education Agency.

Dockweiler, C. J., & Schielack, J. F. (1989). *Technology inservice for school administrators and school board members.* Austin, TX: Texas Education Agency.

Fey, J. T. (Ed.). (1984). *Computing and mathematics: The impact on secondary school curricula.* Reston, VA: National Council of Teachers of Mathematics.

Friel, S., & Joyner, J. (1991). *TEACH-STAT: A project to improve elementary school statistics education.* Proposal funded by the National Science Foundation.

Hall, G. E. (1976). *Longitudinal and cross-sectional studies of the concerns of users of team teaching in the elementary school and instructional modules at the college level* (Research and Development Report No. 3035). Austin, TX: Research and Development Center for Teacher Education. (ERIC Document Reproduction Services No. ED 251 428)

Hall, G. E., & George, A. A. (1979). *Stages of concern about the innovation: The concept, initial verification and some implications.* Austin, TX: Research and Development Center for Teacher Education.

Hall, G. E., Wallace, R. C., & Dossett, W. A. (1973). *A developmental conceptualization of the adoption process within educational institutions* (Research and Development Report No. 3006). Austin, TX: Research and Development Center for Teacher Education. (ERIC Document Reproduction Services No. ED 095 126)

Harvey, J. G. (1991). *Report on the Teaching Mathematics with Calculators project.* Paper presented at the annual research presession of the National Council of Teachers of Mathematics and the Special Interest Group/Research in Mathematics Education, New Orleans, LA.

Hirschhorn, D. B., & Senk, S. (1992). Calculators in the UCSMP curriculum for grades 7 and 8. In J. T. Fey & C. R. Hirsch (Eds.), *Calculators in mathematics education: 1992 yearbook* (pp. 79-90). Reston, VA: National Council of Teachers of Mathematics.

Lamphere, P. M., Usnick, V. E., & Bright, G. W. (1989). *Technology inservice for teachers of mathematics.* Austin, TX: Texas Education Agency.

Morin, K. (1986). The classroom teacher and curriculum developer: A sharing relationship. In K. Zumwalt (Ed.), *Improving teaching* (pp. 149-168). Alexandria, VA: Association for Supervision and Curriculum Development.

Osborne, A. (1991). *Report on the C^2PC project.* Paper presented at the annual research presession of the National Council of Teachers of Mathematics and the Special Interest Group/Research in Mathematics Education, New Orleans, LA.

Prokosch, N. E., Bright, G. W., & Freiberg, H. J. (1991, April). *Classroom observations in middle school mathematics taught with technology.* Paper presented at the annual meeting of the American Educational Research Association, Chicago, IL.

Randall, R. S., & Bennett, J. (1991, July). *An evaluation of the Texas Mathematics Staff Development Program.* Austin, TX: Texas Education Agency.

Resnick, L. B. (1987). *Education and learning to think.* Washington, DC: National Academy Press.
Salomon, G., & Gardner, H. (1986). The computer as educator: Lessons from television research. *Educational Researcher, 15*(1), 13-19.
Super, D. B. (1992). Implementing calculators in a district mathematics program: Three vignettes. In J. T. Fey & C. R. Hirsch (Eds.), *Calculators in mathematics education: 1992 yearbook* (pp. 208-216). Reston, VA: National Council of Teachers of Mathematics.
Young, J. (1985). Participation in curriculum development: An inquiry into the responses of teachers. *Curriculum Inquiry, 15,* 387-414.

Using the Calculator to Develop Number and Operation Sense, K - Grade 5

Jane F. Schielack
Clarence J. Dockweiler
Texas A&M University

Calculators have the potential to transform school mathematics from a procedurally dominated subject to the exciting study of patterns and relationships. (Wheatley & Shumway, 1992, p. 1)

The introductory quote includes the often expressed potential that the calculator, as a tool, provides for transforming school mathematics into something more exciting and valuable than it traditionally has been. This very current comment follows after a similar statement in the *Curriculum and Evaluation Standards for School Mathematics* (National Council of Teachers of Mathematics [NCTM], 1989). The NCTM standards in the K-4 section of the document indicate that "calculators enable children to explore number ideas and patterns, to have valuable concept-development experiences, to focus on problem-solving processes, and to investigate realistic applications" (p. 19). We will attempt to show how this potential of the calculator can indeed enable children to become actively involved in the meaningful explorations of patterns in number and operations.

As the title of this paper indicates, the focus is on the use of calculators with younger children. Concerns are still expressed regarding the use of calculators with younger children. Payne (1990) takes a rather strong stand when he suggests that all children, even those in preschool and kindergarten, should have access to calculators. He extends beyond the availability comment by suggesting that research has produced "no evidence at all that they do any harm" (p. 12). The lack of evidence may not be surprising to the interested researcher, since little research has been conducted with younger learners!

There are those who recognize that the use of calculators can lead to some very positive, and maybe unexpected, results. For example, Worth (1990) points out that children will learn that they must do the thinking and that they must tell the calculator what to do. In addition, she suggests that this required thinking on the part of the learner will lead to the children's use of their mental calculators, which are always available. Along the same lines, Van de Walle (1990) recognizes the importance of facilitating the child's connecting a number counting activity with the numeral. This connection is enhanced when the calculator is the vehicle for displaying the numeral.

These comments set the tone for what we consider to be the thrust of this paper: the need to provide for children activities that incorporate the calculator as a tool to encourage thinking about mathematics by exposing children to patterns and relationships. As a result of these experiences, children will realize the power of their mental calculator and the appropriate circumstances in which to use it.

Goals for the Use of Calculators in the Mathematics Classroom

Before we introduce a few activities to illustrate this purpose, we take the opportunity to consider the goals of calculator use in a mathematics classroom and the research findings that relate to our considerations. The establishment of goals for calculator use in the classroom must, obviously, be found within the context of the general goals for a mathematics classroom. Fuson (1992), in dealing with the research on addition and subtraction of whole numbers, provides interesting insights into the learning of mathematics related to calculator use. She considers the future of our present learners and suggests that the future will create mathematical needs impossible to envision now. She further claims that our envisioned school mathematics classrooms are places where:

> (a) children engage in mathematical situations that are meaningful and interesting to them, (b) the emphasis is on sustained engagement in mathematical situations, not on rapidly obtaining answers, (c) alternative solution procedures are accepted, discussed, and justified, and (d) errors are just expected way stations on the road to solutions and should be analyzed in order to increase everyone's understanding. (p. 268)

Each of the components of Fuson's envisioned classroom includes those factors which we claim would be facilitated by the use of calculators. For example, the use of calculators permits more sustained engagement with a problem than paper-pencil work does, alternative solution procedures are more likely to be accepted procedure with the use of calculators, and so on. The goals enumerated compare favorably with other models, the significance here being that the goals for a mathematics classroom almost require the use of calculators if the goals are to be met.

Research Findings

Current mathematics education literature abounds with discussions of the constructivist approach to teaching mathematics. Much of what has been done in mathematics classrooms in the name of mathematics education reform can be interpreted as constructivist. Schoenfeld (1992) has identified four fundamental assumptions that he suggests are underpinnings of current cognitive research on children's learning of mathematics. His assumptions are that (a) learners construct their mathematical knowledge, (b) instruction should be organized to facilitate this construction, (c) the learner's development of concepts/ideas should provide sequencing information, and (d) skills should be taught in relation to understanding and problem solving.

It is interesting to look at the research on calculator use from the perspective of these four assumptions. It appears that, even though the researcher did not consciously identify with the constructivist camp, the studies have a definite constructivist leaning. Sugarman (1992) concludes from his work "that the change toward a more concept-based approach to the school mathematics curriculum, already overdue but made more urgent by the arrival of the inexpensive pocket calculator, can be effected by a constructivist approach to teaching methodology" (p. 55).

In what appears to be a constructivist approach to learning, Shuard (1992) reports on her efforts in England to develop a Calculator-Aware Number (CAN) curriculum.

Although little hard statistical data were gathered, the study was well thought out, and the results with these young children (six years at the beginning of a five-year project) are fascinating. Initially, each child was provided with a calculator, and the teachers were to allow the children to decide on using or not using the calculator. The teachers were also instructed not to teach the traditional operational algorithms, since it was apparent that the children would not need the algorithms because of calculator availability.

As time progressed, the children became familiar with numbers and with the operations. The students soon developed mental methods for these operations since they did not have written algorithms to employ and the calculators seemed unnecessary. Shuard's report includes fascinating accounts of child-developed approaches to computations with larger numbers in which they rationalized their use of the calculator.

A test "of understanding not only of number but of a variety of mathematical ideas" (Shuard, 1992, p. 43) was administered after one and two years of involvement. This quantitative evaluation and the qualitative evaluation (which the researchers considered more important) resulted in conclusions in favor of the experimental group regarding children's enthusiasm, their understanding of topics usually thought too difficult for their age, their ability in mental calculation, and their persistence.

The oft-cited meta-analysis by Hembree and Dessart (1986) has been extended by those researchers to include an additional nine studies from the late 1980s. Their conclusions follow on from their previous research when they suggest that there is overwhelming evidence to support the idea that calculator use for instructing and testing improves learning and performance with concepts, skills, problem solving, and attitudes (Hembree & Dessart, 1992).

There are, of course, cautions by some researchers. Branca, Breedlove, and King (1992) point out that there are some mathematical topics for which the use of calculators is not appropriate in concept development. They use the example that understanding of fraction concepts and fraction operations is more easily attained through the use of concrete materials. However, research on using the calculator in conjunction with physical representations of rational numbers, similar to that of Shuard (1992) in the CAN curriculum with primary students and whole number concepts, has yet to be done.

Activities to Enhance Concept Development in Young Learners

As Reys (1989) infers in her discussion of the calculator as a tool for instruction and learning, teachers and students can use the calculator to develop topics in new ways, perhaps by generating data and exploring patterns and problem-solving strategies in order to focus on conceptual understanding and critical thinking. In addition, calculator use by young children should always promote a "healthy attitude" toward the calculator as an extension of, not a replacement for, what one knows and understands. The following activities are examples of instruction that combine calculator usage with manipulating concrete models, investigating patterns, and generating and testing hypotheses in order to enhance students' conceptual understanding of number and operations.

Developing Number Sense

According to the *Curriculum and Evaluation Standards for School Mathematics* (NCTM, 1989), "children with good number sense (1) have well-understood number meanings, (2) have developed multiple relationships among numbers, (3) recognize the relative magnitudes of numbers, [and] (4) know the relative effect of operating on numbers" (p. 38). These characteristics come about through the exploration of quantities and their accompanying symbolic representations. In her discussion of the use of calculators to develop number sense, Howden (1989) observes that "activities as simple as successively adding or subtracting ones on a calculator help to develop the concept of order" (p. 8). This idea can be strengthened by having students investigate the connections between the calculator's symbolic representations and accompanying physical representations of whole numbers, fractions, and decimals. Once strong connections between the quantities and symbols are established, the calculator can be used to generate data from which students can identify and analyze patterns and make generalizations and predictions.

Making connections. Students using calculators with a constant function capability can prepare them to count by entering "+ 1". As the teacher places coins or counters on the overhead, one by one, students press "=" on the calculator each time an object is placed. Students should say each number, seeing it pictorially on the overhead screen and symbolically on the calculator.

An auditory component can be included by dropping the coins or counters into a jar as students record on their calculators. The objects from the jar can then be placed on the overhead for students to be able to match the quantity to the symbol. Older students, using a centimeter cube to represent hundredths of a hundreds square, could enter "0.01 = = =" as cubes are placed on the overhead to investigate the connections between the amounts, their symbols, and their relationships to one another and one.

To help move students toward the optimal strategies that are afforded by the use of place value, the calculator can be introduced as a device for recording numbers represented by place-value blocks. As one student places one hundreds square, three tens, and two units on the desk or place-value board, the student's partner can use the constant-function capability of the calculator to "count":

Press: + 100 =   +10 =   =   =   +1 =   =
Say: 100   110   120   130   131   132

Students who have already mastered the standard pattern for reading place-value materials can use the calculator to investigate other methods of combining the hundreds, tens, and ones and make a poster or collage of "The Many Disguises of 132." Older students can use the calculator in the same way with decimal place-value blocks or strips to investigate concrete and symbolic representations of tenths and hundredths.

Making generalizations and predictions. Students can enter "1" on the calculator, discuss its value, and represent the quantity concretely with place-value materials, followed by their entering "2" without clearing the calculator and discussing what has happened to the "1" as they display the new quantity with concrete materials. As students continue to enter digits (without clearing the calculator) and adjust their physical representation of the quantity displayed, they are led to a generalization about what happens to the value of each digit as a new one is entered into the calculator.

Older students should continue this investigation by clearing the calculator, entering "0.1", discussing its value, and displaying it concretely. As successive digits are entered into the calculator and concrete representations are made, students will notice that place value is not determined by position on the calculator, but by relationship to the decimal point. Currently, calculators are being marketed that retain the 0 in the one's place in the display of a decimal less than one and continuously display the decimal point after the one's digit of an integer. These design features make the calculator an even more valid and valuable instructional tool for investigating number.

Developing Operation Sense

> Building an awareness of models and the properties of an operation, seeing relationships among operations, and acquiring insight into the effects of an operation on a pair of numbers ... are aspects of *operation sense*. ... Operation sense interacts with number sense and enables students to make thoughtful decisions about the reasonableness of results. Furthermore, operation sense provides a framework for the conceptual development of mental and written computational procedures. (NCTM, 1989, p. 41)

The calculator can serve as a useful tool for developing operation sense when incorporated in activities that emphasize connections between the many kinds of physical situations associated with an operation and the operation in its symbolic form. Also, young students can broaden their understandings and develop important new understandings of an operation by using the calculator to generate data in activities designed "to explore patterns and relationships for an operation, between operations, and between an operation and other mathematical topics" (Trafton & Zawojewski, 1990, p. 19).

Making connections. Young students can be presented with models of physical settings within which to generate their own stories involving mathematical operations. Some examples of these settings are animals on a farm, shells on a seashore, frogs and fish in a pond, or toys on a shelf. As one partner uses objects to demonstrate and describe the action in the story, the other partner can record the relevant numbers and operations on the calculator.

As they recognize the connections of the physical representations of the numbers and operations to what is being keyed into, and displayed on, the calculator, students begin to build a sense of the calculator's abilities and limitations while developing a healthy trust in their own understanding and control over the calculator. When the sum displayed on the calculator fails to match the total represented by the objects, students question their procedure with the calculator. This experience paves the way for later use of the calculator in computation, emphasizing the value of the calculator to deal with more complicated numbers when accompanied by attention to the procedures used and the reasonableness of the results obtained.

Older students can use place-value materials and the calculator to strengthen estimation skills. Before learning standard algorithms, students could explore multiplication of larger numbers on the calculator. After obtaining a product, for example 156 x 3, with the calculator, students could be encouraged to use visualization to determine the reasonableness of the result by viewing one hundred, five tens, and six ones and imagining three groups of that same size. Questions such as, "How many hundreds would there be? Could you make any new hundreds? How many?" would lead to appropriate estimation strategies, and, possibly, interesting student-generated mental algorithms (e.g., Shuard, 1992).

Making generalizations and predictions. After developing important physical referents for the operations, students can use the calculator to explore the mathematical properties of and relationships between the operations. Students can generate patterns by repeatedly adding two or three or five to various starting numbers and recording these patterns on a hundreds chart. For example, starting at 4 and adding 3 over and over gives a similar pattern to starting at 10 and adding 3 repeatedly. Students then can try to duplicate a particular pattern with repeated subtraction. For example, the previous addition pattern can be generated by starting at 100 and subtracting 3 repeatedly. Questions that have to be answered include, "What amount do I need to subtract each time? Where should I start? Are there any other starting places that would give me the same pattern?" With appropriate direction from the teacher, generalizations about the characteristics of the matching addition and subtraction patterns can be made, with inverse relations and properties of numbers appearing as important ideas.

Factors and multiples, of course, can be investigated with the use of the calculator. In addition, data can be collected to investigate patterns of remainders. If using a calculator with integer division capability, students can consider the following questions: When using a certain divisor, what kinds of remainders occur? What patterns do I see if I organize the dividends according to the remainder I get? How do the remainders change if I use a different divisor? Do any patterns occur when I organize the numbers according to those remainders? While reinforcing basic division facts and the meanings of quotient and remainder, students are also strengthening their skills at organizing and interpreting data that leads to some basic concepts of modular systems.

There are several "classic" problems in which the calculator could be particularly useful in directing student thinking toward the identification and analysis of patterns and the generation of hypotheses of mathematical relationships and properties. Whitin (1989) told a fifth-grade class that the difference between a four-digit number and the number formed by reversing the digits is always 6174. As the students generated examples to investigate the statement, they found that it was true only for certain numbers. The problem then became a search for the common characteristics of the numbers for which the statement held true. Bledsoe (1989) had students investigate the problem of arranging the nine digits 1-9 in a three-by-three array such that the sum of the top two three-digit numbers was the bottom three-digit number. At the heart of the activity was a search for a generalization that could predict which arrangements of numbers would work. Neither of these authors directly mentioned the use of the calculator and even considered these activities as motivating ways for students to practice computation. However, by choosing to use the calculator to generate data for each of these problems, students could quickly generate enough data to find meaningful patterns and devote more time to analyzing the patterns and thinking mathematically about the operations.

Although there are some concerns about using the calculator to develop fraction concepts (e.g. Branca et al., 1992), once students have explored concrete representations of fractions and operating on fractions, the calculator can be used to investigate further the behavior of addition, subtraction, multiplication, and division with fractions. Given four distinct, non-zero whole numbers--a, b, c, d--students can place one number in each of the boxes in the template to generate the largest or smallest sum or difference (or with middle-grade students, product or quotient).

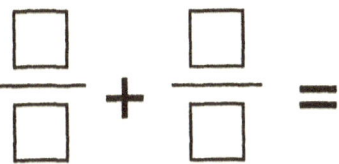

Using a calculator that has the capability to operate on rational numbers in fraction form, students can test their conjectures for placement of the numbers. The calculator allows the investigation of operating on fractions without needing to employ a standard algorithm. Students can repeat the activity with several different sets of numbers, looking for a pattern that will predict how any given set of numbers should be arranged to give, say, the smallest difference. The investigation could be extended to include sets of numbers that include 0 or two of the same number.

Conclusions

From our review of the literature, it appears that little research has been conducted on the integration of the calculator directly into the learning process. The activities we have included are the type that we think would appropriately do just that. We would encourage studies that look at serious attempts to integrate use of the calculator into the learning process.

Although the context is different, we refer again to comments by Fuson (1992), in which she is describing some research difficulties with the topic of addition and subtraction of whole numbers. We take the liberty to paraphrase her conclusion, apply it to the use of calculators, and raise a final note of concern when we suggest that most of our present knowledge about U. S. children's use of calculators is based on research done with children who have received traditional school mathematics instruction. We need more knowledge about how much or in what ways children can learn about mathematics with calculators if they have experiences in the classroom that integrate the use of the calculator into the learning process.

References

Bledsoe, G. J. (1989). Hook your students on problem solving. *Arithmetic Teacher*, 37(4), 16-20.

Branca, N. A., Breedlove, B. A., & King, B. (1992). Calculators in the middle grades: Access to rich mathematics. In J. T. Fey (Ed.), *Calculators in mathematics education: 1992 yearbook* (pp. 9-13). Reston, VA: National Council of Teachers of Mathematics.

Fuson, K. C. (1992). Research on whole number addition and subtraction. In D. A. Grouws (Ed.), *Handbook of research on mathematics teaching and learning* (pp. 243-275). New York: Macmillan.

Hembree, R., & Dessart, D. J. (1986). Effects of hand-held calculators in precollege mathematics education: A meta- analysis. *Journal for Research in Mathematics Education, 17*(2), 83-99.

Hembree, R., & Dessart, D. J. (1992). Research on calculators in mathematics education. In J. T. Fey (Ed.), *Calculators in mathematics education: 1992 yearbook* (pp. 23-32). Reston, VA: National Council of Teachers of Mathematics.

Howden, H. (1989). Teaching number sense. *Arithmetic Teacher, 36*(6), 6-11.

National Council of Teachers of Mathematics. (1989). *Curriculum and evaluation standards of school mathematics.* Reston, VA: Author.

Payne, J. N. (1990). New directions in mathematics education. In J. N. Payne (Ed.), *Mathematics for the young child* (pp. 1-16). Reston, VA: National Council of Teachers of Mathematics.

Reys, B. J. (1989). The calculator as a tool for instruction and learning. In P. R. Trafton (Ed.), *New directions for elementary school mathematics: 1989 yearbook* (pp. 168-173). Reston, VA: National Council of Teachers of Mathematics.

Schoenfeld, A. H. (1992). Learning to think mathematically: Problem solving, metacognition, and sense making in mathematics. In D. A. Grouws (Ed.), *Handbook of research on mathematics teaching and learning* (pp. 334-370). New York: Macmillan.

Shuard, H. (1992). CAN: Calculator use in the primary grades in England and Wales. In J. T. Fey (Ed.), *Calculators in mathematics education: 1992 yearbook* (pp. 33-45). Reston, VA: National Council of Teachers of Mathematics.

Sugarman, I. (1992). A constructivist approach to developing early calculating abilities. In J. T. Fey (Ed.), *Calculators in mathematics education: 1992 Yearbook* (pp. 46-55). Reston, VA: National Council of Teachers of Mathematics.

Trafton, P. R., & Zawojewski, J. S. (1990). Implementing the *Standards*: Meanings of operations. *Arithmetic Teacher, 38*(3), 18-22.

Van de Walle, J. A. (1990). Concepts of number. In J. N. Payne (Ed.), *Mathematics for the young child* (pp. 63-88). Reston, VA: National Council of Teachers of Mathematics.

Wheatley, G. H., & Shumway, R. (1992). The potential for calculators to transform elementary school mathematics. In J. T. Fey (Ed.), *Calculators in mathematics education: 1992 yearbook* (pp. 1-8). Reston, VA: National Council of Teachers of Mathematics.

Whitin, D. J. (1989). The power of mathematical investigations. In P. R. Trafton (Ed.), *New directions for elementary school mathematics: 1989 yearbook* (pp. 183-190). Reston, VA: National Council of Teachers of Mathematics.

Worth, J. (1990). Developing problem-solving abilities and attitudes. In J. N. Payne (Ed.), *Mathematics for the young child* (pp. 39-62). Reston, VA: National Council of Teachers of Mathematics.

The Calculator Project: Assessing School-Wide Impact of Calculator Integration

Gary G. Bitter
Mary M. Hatfield
Arizona State University

The calculator has become a ubiquitous and indispensable tool in American culture. As such, instruction in its use should be a regular part of the mathematics curriculum in our schools at all grade levels (National Council of Teachers of Mathematics, 1989). Two basic concerns regarding classroom calculator use have been that (a) calculator use could be detrimental to students' basic mathematics skills and (b) calculator use could make it difficult to test students' mathematics skills.

Research shows that students at all grade levels can benefit from an integration of calculators into the curriculum. Hembree and Dessart (1986) conducted a meta-analysis of 79 calculator studies and concluded that in about half of the studies, student achievement was higher for the calculator group, and in about half of the other studies, the calculator group had achievement scores that were as good as students using traditional methods. (One exception was at the fourth-grade level, where a decline in students' scores was reported when the calculator was used.) Student attitude toward mathematics and student self-concept with respect to learning mathematics were significantly better for the calculator group than for the non-calculator group. Another review of almost 200 research studies (Suydam, 1990) indicated that learning basic facts, counting skills, computation with whole numbers, fractions and decimals, and estimation skills were enhanced by using calculators.

Research findings indicate that, contrary to the fears of traditionalists, calculators can greatly assist learning and do not diminish students' ability to perform computation skills. More specifically, research sheds light on how the use of calculators affects achievement, the mastery of skills, such as computation with whole numbers, and attitudes toward mathematics.

Problem solving performance has been found to be significantly better for students who use calculators (Szetela & Super, 1987). When calculators were used in problem solving activities, students changed their focus from computation to problem solving strategies and processes. According to Suydam (1982), calculators will aid students in developing problem solving strategies when the focus of the problem is shifted from computation to the process of solving the problem.

The National Council of Teachers (NCTM) in its *Curriculum and Evaluation Standards for School Mathematics* (NCTM, 1989) states that students should use calculators in appropriate computational situations and that calculators should be viewed as powerful problem solving tools. The *Standards* advocates that students should "select and use an appropriate method for computing from among mental arithmetic, paper-and-pencil, calculator, and computer" (p. 94).

As students actively engage in learning mathematical concepts, their confidence in their ability to succeed in mathematics increases. The gender differences in mathematics achievement that appear at the middle-school level (Lappan, Reyes, & Stanic, 1988) can also be minimized when females have confidence in their abilities to do mathematics and exhibit achievement-related behavior and attitude (Hart & Stanic, 1989).

The appropriate integration of calculators into mathematics instruction can enrich the current mathematics curriculum. Estimation and mental math skills take on increased importance with extensive calculator use, and many researchers believe that mental calculations and estimation are two of the best means of developing a child's understanding of numbers and their properties (Hope & Sherrill, 1987). Students must develop a feel for the reasonableness of an answer, and they need to incorporate estimation as a regular part of the problem solving process (Dick, 1988).

In view of the evidence of the effectiveness of calculators in mathematics instruction and given the endorsement of virtually every mathematics education organization, a question arises as to why school personnel have not more greatly supported classroom calculator use. Some possible reasons follow:
1. Teachers are habituated to a paper-and-pencil curriculum.
2. School personnel are unaware of or not convinced of the potential benefits of using calculators in mathematics instruction.
3. School personnel are concerned about how calculator use will affect standardized test scores.
4. Existing mathematics curricula do not offer an integrated approach to using calculators in mathematics instruction.
5. School personnel are uncertain how to implement a calculator-enhanced mathematics program.
6. School personnel are concerned that all students do not have equal access to a calculator.

We need more experimental and experiential knowledge to explore these possibilities and to indicate how best to implement and integrate calculators into the mathematics curriculum. To address these needs, a study was conducted at a middle school in Arizona.

The major focus of the study was the effect of calculator use on standardized mathematics test scores. Because few standardized tests currently allow calculators, some educators fear that increased use of calculators in the classroom will result in lower standardized test scores. This study directly investigated the question: How do students perform on a standardized mathematics test when they use a calculator, compared with when they do not? The study also considered these secondary research questions:
1. Do calculators affect girls' achievement differently from boys' achievement?
2. How do students, parents, and teachers feel about a major calculator implementation effort and what is the state of calculator proliferation among these groups?
3. How does the degree to which a teacher who integrates calculators in a systematic, comprehensive way, affect the achievement and attitudes of students?
4. What is the relationship between student attitude toward mathematics and achievement using calculators?

Method

Subjects

Subjects were 501 seventh and eighth-grade students from a middle school located in the suburbs of a large southwestern city; all subjects were issued Texas Instruments Math Explorer calculators for the school year. Fifty-four percent of the subjects were seventh-graders and 46% were eighth-graders. The gender distribution was 54% male and 46% female. Seventy-four percent of the subjects were White, 10% were Hispanic, 9% were Native American, 2% were African-American, 2% were Asian/Pacific Islander, and 3.9% did not report their ethnicity. The student population at this school is highly transient with a yearly turnover of approximately one-third of the students. One-third of the incoming seventh-graders qualify for Chapter I Services in reading and mathematics.

Data were also collected from 323 of the students' parents. Seven project teachers participated in the study: two teachers from seventh grade, two from eighth grade, and three special education teachers.

Procedures

Instruments. Student mathematics performance was measured on the mathematics subtests of the Iowa Test of Basic Skills (ITBS): concepts, problem solving, and computation. The ITBS is representative of most current standardized tests in that it is not specifically designed to be taken using a calculator, making it suitable for this study, which investigated the relationship between calculator use and standardized test scores.

In addition to the standardized achievement test, five instruments developed by the researchers were administered, three to students, one to parents, and one to teachers. Student data were collected via a scaled mathematics attitude instrument, a calculator attitude and use survey, and an open-response questionnaire. Parents responded to a parent calculator inventory that included use and attitude toward calculators, while teachers responded to a calculator attitude questionnaire.

The student attitude questionnaire consisted of 21 items on a four-point Likert scale (strongly agree to strongly disagree) measuring attitudes toward calculators and a scaled instrument measuring attitudes toward mathematics (Dutton & Adams, 1961). On the latter, students checked the statements that applied to their feelings about mathematics and indicated their general mathematics attitudes along an eleven-point continuum scale ranging from strongly favor to strongly against. The two scales of attitude toward mathematics correlated significantly ($r=.68$, $p<.01$).

The student calculator survey measured how and when students used calculators, both in the home and at school. Students checked items that indicated ways they had used the calculator, and reported the average amount of time spent using the calculator in and out of the classroom. The survey instrument included two general attitude items about using the calculator for mathematics. The first measured how students felt, and the second measured how students perceived their parents felt. Each attitude item was measured on a five-point Likert scale that ranged from very positive to very negative.

An open-ended questionnaire asked students to respond freely to six questions regarding their feelings about using calculators. The qualitative dimension of the study was included to provide another perspective when viewing the quantitative data.

The teacher attitude scale, a revised version of the Pocket Calculator Attitude Scale (Bitter, 1980), assessed teachers' attitudes toward calculator use in general and toward students' calculator use for learning mathematics. The 20-item questionnaire used a five-point Likert scale that ranged from strongly agree to strongly disagree.

Parents were surveyed to determine their use of calculators and their attitudes toward using calculators themselves and toward their child's use of calculators. The attitude section contained nine items measured along a four-point Likert scale from strongly agree to strongly disagree.

Teacher inservice. Teacher inservice training was on-going throughout the two years of the project. During the pilot year participating teachers and the school principal attended weekly hour-long inservice sessions before school. During the project's implementation year (the second year of the project), in addition to their individual planning times, the teachers were assigned a common planning time during the school day where they could interact and collaborate.

During inservice training teachers worked through calculator lessons in a problem solving environment in which they participated as learners. Sample lessons integrating the calculator into the existing mathematics curriculum were developed and modeled. The mathematical learning experiences were constructed to engage teachers in a direct and dynamic way; teachers developed activities, exchanged ideas, shared teaching strategies, and discussed problems. A variety of instructional formats were used including cooperative learning, peer tutoring, and small and whole group instruction. Teachers were encouraged to request the staff to provide specific training on mathematics topics of interest or concern to them.

As a component of the long-term support, project staff observed the teachers regularly and provided feedback for improving instruction. The project staff was available to teach, providing the opportunity for teachers to visit other classrooms and observe other teaching techniques. Frequent discussions between project staff and teachers were aimed at encouraging teachers to engage in the process of self-reflection and self-analysis in an effort to improve instruction.

Data collection. Results based on the data gathered during the pilot year were used to finalize the research design and data collection instruments. In year two, students took the Iowa Test of Basic Skills mathematics test on two occasions: during the week of April 2-6, 1990, students took the test without using calculators; during the week of April 30-May 4, 1990, students took a parallel form of the test using calculators. Parallel forms were used to help obviate any practice effect, as was the three week interim between test administrations, an interval considered sufficient for viable retesting (Cronbach, 1984). (State law required that all students take the official, non-calculator administration of the ITBS before any "experimental" administrations, thus prohibiting a counterbalanced design.) All other instruments were administered in early September, 1989 and in mid-May, 1990 (the beginning and end of year two).

Observational data were gathered regarding how teachers integrated calculators into their curriculum. The project research assistant, who is also an experienced teacher, observed teachers as they taught their mathematics classes and recorded teacher and student activities. Semi-structured interviews of students and teachers examined their attitudes about using the calculator for mathematics and their descriptions of the calculator's role in the teaching and learning of mathematics. These data were used to assess the degree to which participating teachers adhered to suggestions about calculator use and to the implementation of guidelines described by NCTM (1989).

Data Analysis: Qualitative teacher data

Based on interview and observational data, project researchers assessed the degree to which each project teacher used calculators in a consistent, systematic manner. The NCTM *Standards* served as the framework for evaluating "appropriate" calculator use (meaning calculators were used to enhance the mathematics or to eliminate tedious computations, rather than to check homework or to compute grades). In spite of the project's on-going inservice training, analysis of the teachers' instructional activities provided evidence of minimal calculator use, except for one eighth-grade teacher (called here "Ms. Spanos"). The following description of Ms. Spanos illustrates characteristics judged to be in conformance with NCTM guidelines.

In general, Ms. Spanos followed sound instructional practice; she began class with a stated instructional objective, then organized the class into appropriate size cooperative groups. She emphasized process learning, used an inquiry approach that invited student questions, and often asked students to explain their reasoning and to justify their answers. Ms. Spanos circulated about the classroom to check for student understanding. Most of her lessons involved engaging students in active ways (using models, manipulatives, and games) in doing exploratory work.

With respect to the calculator, Ms. Spanos integrated the calculator in a consistent, comprehensive way across topics of mathematics throughout the year. Her lessons included a broad range of calculator use, such as estimation, problem solving, decimals, fractions, large number computation, ratios, proportion, prime factorization, order of operations, rounding, percent and various mathematical activities (palindromes, magic squares, probability experiments, and so on). She continually focused on having students select the appropriate method for finding answers.

Unexpectedly, Ms. Spanos was the only teacher in the study who integrated calculators according to NCTM guidelines and standard instructional design practices outlined during project development by university personnel. Despite inservice training, the other teachers followed a "traditional" teaching approach, which did not involve a calculator-enhanced mathematics program.

Achievement

Achievement data were analyzed using two multivariate analyses of variance. Because seventh and eighth-graders used different levels of the ITBS tests, separate analyses were performed for each grade. The three mathematics subtest scores for concepts, problem solving, and computation were criterion variables and calculator use (with and without) was a within-subjects factor. Three other factors were considered: sex, teacher, and student attitude toward mathematics. The teacher factor was included to investigate the effect on student achievement of teacher adherence to a systematic, calculator-enhanced mathematics program.

As stated above, none of the seventh-grade teachers could be judged to have integrated calculators according to established project criteria, thus the teacher factor could not be considered in the analysis of seventh-grade scores. Subsequently, it was decided to limit each of the two analyses to two of the three between-subjects factors (sex, teacher, and attitude toward mathematics).

Eighth-grade scores were analyzed using a 2 x 2 x 2 MANOVA where the three ITBS mathematics subtests were criterion variables, calculator use was a within-subjects factor, and sex and teacher were between-subjects factors. The design for eighth-graders consisted of 47 subjects in the largest cell and 44 subjects in the smallest cell. The design for seventh-graders was a 2 x 2 x 2 MANOVA with calculator use as the within-subjects factor, and sex and attitude toward mathematics as between-subjects factors. The seventh-grade design was not initially well-balanced (almost twice as many students fell in to the "positive attitude" cells), so subjects were randomly dropped from some of the cells to produce a balanced design (N=33 for each cell).

In each analysis, if an effect showed multivariate significance, univariate F-tests were examined to isolate the performance effect for each of the three mathematics subtest scores (concepts, problem solving, and computation).

Attitude and Inventory

Attitude and inventory data for students, parents, and teachers were summarized. The percentage of each group responding at each level of the attitude scale (strongly agree through strongly disagree) before and after the project was computed, and student attitude data were summarized for subgroups by sex.

Student responses on the open-ended questionnaire were analyzed qualitatively until categories of responses emerged from the data. Student responses were placed in categories that arose from the data themselves, and the percentage of students responding in each category was computed across all students and for subgroups by sex and ethnic background.

Results

Achievement

Eighth-Graders. Table 1 shows eighth-graders' grade-level equivalent score by calculator, sex, and teacher. Cumulative means for each factor (calculator, sex, and teacher calculator integration) are shown below the main body of the table. The cumulative means for the calculator factor show that students did better using a calculator on all three mathematics subtests. The calculator effect was statistically significant for each subtest ($F(1,180)=16.36$, $p<.001$ for concepts, $F(1,180)=200.43$, $p<.001$ for computation, and $F(1,180)=6.31$, $p<.02$ for problem solving).

Cumulative means by sex do not seem to reveal an overall performance difference between girls and boys, though statistically, girls scored higher on the computation subtest ($F(1,180)=7.59$, $p<.01$). A sex-by-calculator interaction, which showed multivariate and univariate significance, was found and is illustrated in Figure 1. The figure shows that girls improved more than boys on each of the three mathematics subtests. The difference in improvement was significant for problem solving ($F(1,180)=13.01$, $p<.001$) and for computation ($F(1,180)=4.61$, $p<.04$), and approached significance for concepts ($p<.08$). This effect shows that eighth-grade girls improved more than boys using a calculator.

Table 1. *Grade Equivalent Score of Eighth Graders by Calculator Use, Sex, and Calculator Integration*

	Boys				Girls			
	Integrated (N=47)		Nonintegrated (N=47)		Integrated (N=44)		Nonintegrated (N=46)	
Tests	Mean	sd	Mean	sd	Mean	sd	Mean	sd
With Calculator								
Concepts	9.8	1.6	8.5	1.5	9.7	1.8	8.3	1.4
Problem Solving	9.6	1.8	8.1	1.4	9.6	1.5	8.1	1.4
Computation	9.7	1.4	8.5	1.6	10.4	1.0	8.9	1.0
Without Calculator								
Concepts	9.6	1.8	8.3	1.2	9.2	1.7	7.7	1.4
Problem Solving	9.4	1.9	8.6	1.2	9.0	1.4	7.5	1.7
Computation	8.4	1.2	7.4	1.1	8.5	0.9	7.7	0.9

	CUMULATIVE MEANS					
	Calculator		Sex		Calculator Integration	
Concepts	With	9.1	Boys	9.1	Integrated	9.6
	Without	8.7	Girls	8.7	Nonintegrated	8.2
Problem Solving	With	8.8	Boys	8.9	Integrated	9.4
	Without	8.6	Girls	8.5	Nonintegrated	8.1
Computation	With	9.4	Boys	8.5	Integrated	9.2
	Without	8.0	Girls	8.9	Nonintegrated	8.1

The cumulative means for teacher calculator integration in Table 1 also reveal that students whose teacher used the calculator in an integrated, consistent manner outscored other students on all three subtests ($F(1,180)=43.93$, $p<.001$ for concepts, $F(1,180)=40.97$, $p<.001$ for problem solving, and $F(1,180)=60.57$, $p<.001$ for computation). In addition to a significant teacher main effect, analysis revealed a significant multivariate teacher-by-calculator interaction. Subsequent univariate analysis revealed that the difference on the computation subtest was significant ($F(1,180)=6.49$, $p<.02$), while no significant difference was found for the concepts subtest or the problem solving subtest. This effect shows that eighth-grade students of the teacher who integrated the calculator into the mathematics curriculum improved more on computation than students of teachers who did not.

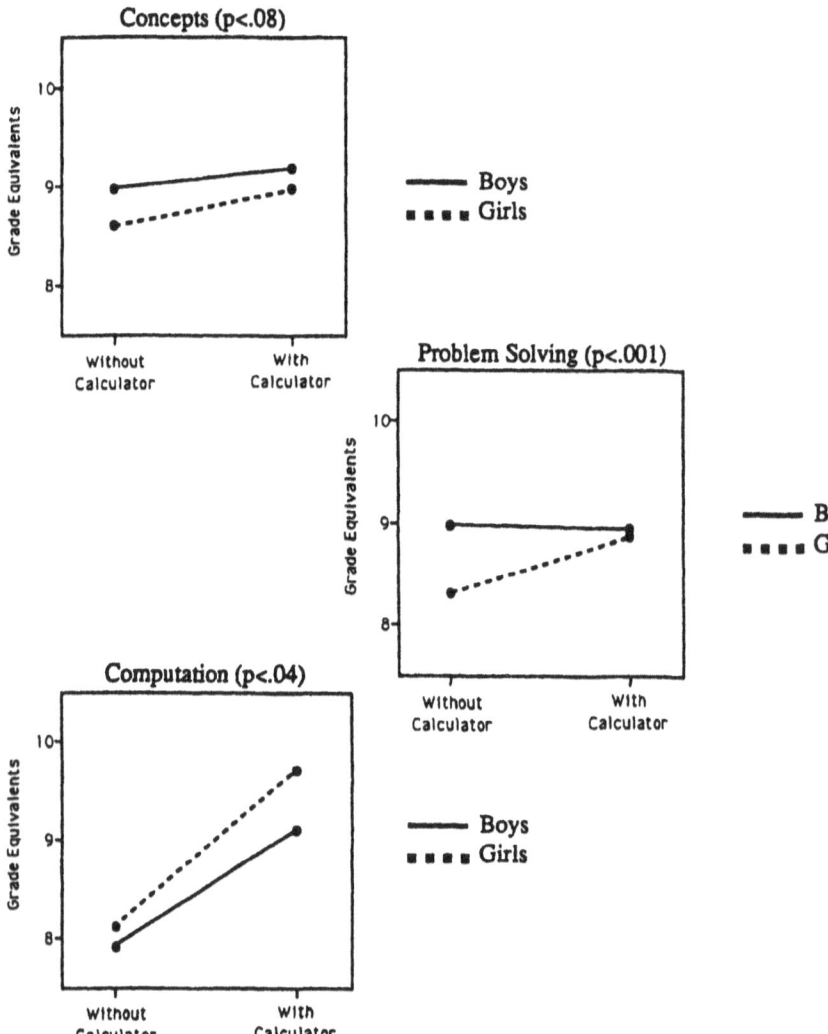

FIGURE 1. *Sex by calculator/non-calculator interaction for eighth-grade student achievement.*

Seventh-Graders. Table 2 shows analogous results for seventh-graders. Statistically, the cumulative means for the calculator factor indicate a significant difference only on computation ($F(1,180)=15.10$, $p<.001$), where student scores improved using a calculator.

Table 2. *Grade Equivalent Score of Seventh Graders by Calculator Use, Sex, and Calculator Integration*

	Boys				Girls			
	Integrated (N=33)		Nonintegrated (N=33)		Integrated (N=33)		Nonintegrated (N=33)	
Tests	Mean	sd	Mean	sd	Mean	sd	Mean	sd
With Calculator								
Concepts	8.2	1.4	7.6	1.6	8.0	1.6	7.9	1.4
Problem Solving	8.0	1.8	7.1	1.9	7.4	1.7	7.8	1.4
Computation	8.1	1.3	7.7	1.6	8.1	1.3	7.8	1.7
Without Calculator								
Concepts	8.4	1.5	7.7	1.4	7.9	1.6	7.6	1.4
Problem Solving	8.0	1.7	7.6	1.7	8.1	1.2	7.3	1.5
Computation	7.5	1.1	7.0	1.3	7.7	1.2	7.5	1.0

	CUMULATIVE MEANS					
	Calculator		Sex		Calculator Integration	
Concepts	With	7.9	Boys	8.0	Integrated	8.1
	Without	7.7	Girls	7.9	Nonintegrated	7.7
Problem Solving	With	7.6	Boys	7.7	Integrated	7.9
	Without	7.8	Girls	7.7	Nonintegrated	7.5
Computation	With	7.9	Boys	7.6	Integrated	7.9
	Without	7.4	Girls	7.8	Nonintegrated	7.5

Unlike eighth-grade scores, seventh-grade scores revealed no statistical differences between boys and girls on the with- and without-calculator tests. That is, the impact of using calculators was the same for boys and girls in the seventh grade. No significant achievement differences were found between students with positive attitudes towards mathematics and students with negative attitudes. Analysis also revealed no significant attitude-by-calculator interaction.

Attitude and inventory

Students. Table 3 shows the percentage of students who responded "Agree" or "Strongly Agree" to items on the scaled student attitude questionnaire before and after the project. The table shows overall percentages before and after as well as percentages for boys, girls, students taught by the teacher who systematically integrated calculators, and students taught by other teachers.

Table 3. *Percent of Students Responding "Agree" or Strongly Agree" to Attitude Items Before (B) and After (A) the Project, Before and After by Sex, and by Teachers' Adherence to Integrating the Calculator into the Mathematics Curriculum*

Attitude Item	N =	Totals B 443	Totals A 488	Boys B 224	Boys A 262	Girls B 219	Girls A 226	NonI 363	Int 124
Students should not be allowed to use a calculator while taking math tests.		30.4	28.3	24.3	21.0	36.7	36.7	21.8	13.7
The calculator will hinder students' understanding of the basic computation skills.		32.9	36.6	35.2	33.8	30.5	40.0	38.8	30.0
Calculators make mathematics fun.		77.9	79.1	78.9	77.4	77.0	81.1	80.1	75.0
Since I will have a calculator, I do not need to learn to do computations on paper.		14.7	12.7	15.9	15.5	13.5	9.4	13.9	9.8
Mathematics is easier if a calculator is used to solve problems.		77.2	86.3	78.1	87.0	76.3	85.4	85.4	88.7
I understand mathematics better if I solve problems with paper and pencil.		42.1	56.2	40.6	56.8	43.6	55.5	57.5	52.5
I know how to use a calculator very well.		73.0	87.6	74.3	87.3	71.7	88.1	89.0	83.6
It is important that everyone learn how to use a calculator.		78.3	85.6	81.5	85.4	75.0	85.8	85.9	84.5
I would do better in math if I could use a calculator.		68.1	72.7	72.7	72.1	63.4	73.4	75.0	65.0
I prefer working word problems with a calculator.		63.8	69.6	66.0	68.7	61.6	70.6	71.5	63.9
I would try harder in math if I had a calculator to use.		54.3	49.3	53.7	50.6	54.9	47.8	50.8	43.4
Using a calculator to solve money problems is confusing.		15.6	12.4	15.9	13.6	15.4	11.1	12.5	12.2
Calculators should be used only to check my answers once I have worked the problems with pencil and paper.		28.7	30.0	30.8	32.5	26.5	27.1	30.3	29.3
Calculators are not useful for solving fraction problems.		21.5	19.6	19.0	12.5	24.2	27.9	9.2	7.4
I feel calculators should not be used on math homework.		13.2	11.9	13.8	14.8	12.7	8.4	12.6	10.7
I am good in mathematics.		67.5	67.9	63.5	74.2	71.7	60.6	66.1	73.8
Using a calculator in math will cause me to forget how to do basic computation skills.		17.3	22.1	19.1	23.6	15.5	20.5	24.8	13.3
I would appreciate math better if I had a calculator to use.		61.6	61.4	65.1	61.2	58.0	61.6	62.4	57.5
I would do better in problem solving if I could use a calculator.		64.8	70.2	68.5	69.6	60.9	70.9	70.9	66.7
If I use a calculator, my estimation skills will decrease.		13.4	26.1	16.3	29.2	10.5	22.5	29.7	14.3
Mathematics is boring.		31.5	38.6	30.8	34.1	32.3	43.8	39.8	33.3

Notes. B=Before; A=After; NonI=NonIntegrated; Int=Integrated

Table 4 shows results from the open-ended questionnaire for all students. These data suggest two trends. Girls apparently felt more "empowered" than boys when using a calculator (by approximately 10% in questions 1, 4, and 6). The term "empowerment" emerged as the category name for responses related to students' feelings of self-efficacy and confidence. The notion of mathematical power is described in the *Curriculum and Evaluation Standards for School Mathematics* (NCTM, 1989), as the development of higher-order processing capabilities and self-confidence. Girls, overall, reported feeling more confident and smarter when they could use a calculator, less confident and more likely to fail when they could not use a calculator.

Table 4. *Percentage of Students by Sex and Who Responded in Particular Categories on Open-Ended Questionnaire*

Items and Student Responses	Total N=398	Boys N=191	Girls N=207
1. When I use the calculator in math class I feel:			
Empowered (smarter, confident, understand)	28.6	23.6	33.3
Better grades (better grades, do better)	12.6	14.7	10.6
General positive (fine, good, happy)	17.8	23.0	13.0
Useful tools (faster, functional, easier)	14.3	13.6	15.0
General negative (bored, dumb, stupid, cheating)	7.0	7.9	6.3
Neutral (no smarter, nothing, no answer)	19.6	18.3	20.8
2. The calculator should be used to do:			
Restricted (specific topics, check work)	80.7	78.5	82.6
Unrestricted (everything, all the time)	17.6	19.4	15.9
Neutral (don't care, don't know, no answer)	1.8	2.1	1.4
3. Using a calculator helped me understand:			
Restricted (specific topics, check work)	64.3	61.8	66.7
Useful tool (correct mistakes, calculation)	11.1	11.5	10.6
Unrestricted (everything, all the time)	5.8	6.8	4.8
Neutral (don't care, don't know, no answer)	18.6	19.4	17.9
4. When I was without the calculator in math class, I felt:			
Empowered (smart, better, really learning)	17.8	18.8	16.9
Unempowered (harder, less confident, stupid)	33.7	28.8	38.2
General positive (okay, good)	11.1	10.5	11.6
General negative (deprived, frustrated, bored)	11.8	16.2	7.7
Neutral (no different, nothing, no answer)	25.6	25.7	25.6
5. When I didn't get to use the calculator on tests, I felt:			
Empowered (do better, smart)	15.1	15.2	15.0
Unempowered (would fail, less confident)	31.9	29.8	33.8
General positive (good)	14.6	16.2	13.0
General negative (took longer, bored, cheated)	18.6	19.4	17.9
Neutral (same, don't know, nothing, no answer)	19.8	19.4	20.3
6. If I don't have a calculator to use next year in math, I'll feel:			
Empowered (will learn more, will study harder)	10.6	9.4	11.6
Unempowered (have hard time, might fail)	20.9	16.2	25.1
Buy own tool (will buy my own)	21.6	19.9	23.2
General positive (okay)	4.0	4.7	3.4
General negative (dislike math, will take longer)	19.1	22.5	15.9
Neutral (don't know, no answer)	23.6	27.2	20.3

The second trend was that 80% of the students favored using calculators in a restricted manner rather than in an unrestricted way. Restricted use included using calculators for a specific topic in mathematics or for a particular task (finding percentages for grades or doing mathematics homework), while unrestricted use included using calculators "for everything." (Refer to Table 4, questions 2 and 3.)

Parents. Table 5 shows the percentage of parents who responded "Agree" or "Strongly Agree" to items on the parent attitude questionnaire administered at the beginning and end of their child's participation in the project. The results show that (a) parents were generally positive about calculator use, (b) after the project, nearly 90% of the parents felt the calculator should be a regular part of the curriculum, (c) after the project, only 12% of the parents felt the calculator was not useful for solving fraction problems, and (d) after the project, 40% of the parents felt that calculators would hinder students' understanding of the basic computation skills.

Table 5. *Percentage of Parents Who Responded "Agree" or "Strongly Agree" on Parent Attitude Items Before and After Project Participation*

Statement	Before (N=144)	After (N=129)
The calculator should be a regular part of the curriculum.	81.4	88.0
Students should not be allowed to use the calculator when taking math tests.	44.1	38.3
Calculators can stimulate a child to learn mathematics.	75.5	82.6
The calculator will hinder students' understanding of the basic computation skills.	32.2	40.5
The calculator makes mathematics fun.	90.3	88.9
Calculators are not useful in solving fraction problems.	17.3	11.7
Calculators should be available only to students who are good in basic computation skills.	40.2	37.0

Teachers. Table 6 shows the percentage of project teachers who responded "Agree" or "Strongly Agree" to items on the teacher attitude questionnaire administered at the end of the year. Teachers unanimously agreed that calculators make mathematics fun. Seventy-five percent of the teachers supported the idea that calculators should be a regular part of the school curriculum and approximately 60% of the teachers felt that calculators will not hinder students' understanding of the basic computational skills. Teachers were generally undecided or not agreeable to allowing students to use calculators while taking mathematics tests.

Calculator Ownership and Use. The median value of the number of calculators in the family was "more than 3" according to both students and parents. Fewer than 2% of parents reported no calculators in the home; 97% of parents reported using a calculator at least once a week. Students reported using a calculator, on the average, between 15-20 minutes in class and between 10-15 minutes outside of class each school day.

Table 6. *Percentage of Teachers (and their Principals) Who Agreed or Disagreed on Selected Teacher Attitude Items Before and After Project Participation*

Statement	Before (N=5)	After (N=6)	Response
Calculators make mathematics fun.	100%	100%	Agreed
Calculators are too expensive for classroom use.	100%	83%	Disagreed
I feel that the calculator should be a regular part of the school curriculum.	80%	100%	Agreed
Students should not be allowed to use the calculator while taking math tests.	60%	67%	Disagreed
Calculators will hinder students' understanding of the basic computation skills.	40%	83%	Disagreed
Calculators are causing students to lose the chance to do mental computation in school.	80%	87%	Disagreed
Calculators should be available to students in all grades.	80%	87%	Agreed

Students reported teacher-directed calculator use as well as self-directed use. Table 7 shows the percentage of students who used a calculator in each of several teacher-directed activities. Students reported that calculators were used more for higher-order activities such as mathematical investigations and introducing new concepts (62% and 59%) than for computing grades and checking work (49% and 37%). Table 8 shows the percentage of students who reported using a calculator in each of several self-directed activities.

Table 7. *Use of a Calculator for Particular Teacher-Directed Activities*

Activity	Percentage
For math investigations	62.2
When new concepts were introduced	59.2
To practice previously learned skills	58.9
To compute grades	49.2
To check paper/pencil computations	37.2
Other	15.8

Table 8. *Use of a Calculator for Particular Self-Directed Activities*

Activity	Percentage
To work in the classroom	78.0
To do my math assignments at home	76.6
When I work independently	72.0
When I work in cooperative learning groups	63.1
Other	13.2

Table 9 shows the percentage of students who used a calculator, in school or independently, for each of several mathematics topics. The table shows that (a) a high percentage of students reported using the calculator for topics that take advantage of the unique capabilities of this particular calculator, specifically fractions (88%) and decimals (87%), (b) a majority of students reported using the calculator for problem solving situations, namely word problems (78%), equations (61%), and algebra (55%), and (c) students reported using calculators for a diversity of topics.

Table 9. *Use of a Calculator for Particular Math Topics*

Topic	Percentage
Fractions	88.4
Decimals	86.7
+,-,x, /	86.5
Percent	78.4
Word Problems	72.2
Equations	60.6
Algebra	55.2
Ratio/Proportion	47.1
Geometry	36.2
Measurement	35.7
Probability	34.1
Statistics	33.9

Discussion

The primary purpose of this study was to implement and integrate calculators into a middle school mathematics curriculum and to evaluate the effects on student achievement in mathematics and attitudes of students, parents, and teachers. The study, which was piloted in its first year and fully implemented in its second year, analyzed achievement and attitudinal data, inventoried students' and parents' calculator use, and analyzed qualitative data about instructional practices of participating teachers and students' open-ended responses regarding using calculators in mathematics.

Students' mathematics performance improved significantly when they had access to a calculator. Eighth-grade students improved on all three Iowa Test of Basic Skills (ITBS) mathematics subtests (concepts, problem solving, and computation), while seventh-graders improved on the computation subtest.

Four factors may account for students' improved mathematics performance when using a calculator. First, students may have committed fewer computational errors with a calculator. Second, students may be able to perform computations more quickly with a calculator and thus complete more problems within the allotted time. Third, students may have been able to focus their attention on higher-order thinking skills rather than on computation. Fourth, students may have felt more confident when using a calculator.

Computational efficiency (the first two factors -- computational accuracy and computational speed) is consistent with the significant results found for both grade levels on the ITBS computation subtest. It is not clear to what extent each of the factors above affected eighth-graders' improvement on the concepts and problem solving subtests. In any case, student performance, overall, on this particular standardized test, improved using a calculator. This result should help mitigate the concern of educators that calculator use will result in lower scores on standardized mathematics tests that are not designed for calculators. While no overall performance difference were found between boys and girls at either grade level, a sex-by-calculator interaction effect was found for eighth-graders; perhaps calculators "helped" eighth-grade girls more than they helped eighth-grade boys. However, no sex-by-calculator interaction for achievement was found for seventh-graders, which prompts caution in generalizing the calculator-by-sex effect, and suggests the need for further research.

The teacher's importance is illustrated by the teacher main effect for eighth-graders and by the teacher-by-calculator interaction for eighth-graders on computation. One third of the eighth-grade students were taught by a teacher committed to implementing calculators in mathematics instruction, as confirmed by observational data. This teacher's students performed significantly better on all ITBS mathematics subtest, both with and without the calculator. This is probably the result of superior calculator implementation, although it is possible that Ms. Spanos was simply a superior teacher in every respect. In either case, the result enforces the importance of training teachers how to integrate calculators into their classrooms and exhorts the importance of both inservice and preservice teacher training. The failure of the project's inservice training to induce other teachers to adequately integrate calculator-based activities into the mathematics curriculum suggests the magnitude of effort that may be required to prepare and motivate teachers to use this and other new technologies.

For the sex-by-calculator interaction and the teacher-by-calculator interaction, the magnitude of differences between non-calculator and calculator scores suggests an increase from the concepts subtest to the problem solving subtest to the computation subtest. This supports the *prima facie* notion that conceptual mathematics is least amenable to calculator intervention. One would assume that problem solving might be amenable in that students would be free to concentrate on the higher-order aspects of the problem, instead of the computation. The teacher could devote more class time to problem solving since tedious computation is minimized with a calculator. Computational mathematics, however, is influenced directly by calculator intervention.

Student attitude towards mathematics was addressed in the analysis of seventh-grade scores; performance of students with positive attitudes was compared with performance of students with negative attitudes. While it was somewhat surprising that no performance difference was found between students with positive and negative attitudes, the fact that no interaction effect was found between attitude and calculator suggests that students improved "equally" no matter what their attitude towards mathematics. This "non-effect" intimates that calculators may "help" students with negative attitudes as much as they help students with positive attitudes.

Descriptive data from the calculator inventory indicates that students reported a median value of "more than 3" calculators counting both their home and at school; parents also reported a median value of "more than three" calculators in the home. This finding bears on two issues in calculator use: access and the dependency problem. Access is the concern with what happens if some students do not have a calculator. Project results suggest such concern is unfounded, assuming the type of calculator is not an issue. The second concern is that calculators promote a dependence on technology that would be detrimental if students found themselves suddenly bereft of calculators. The results indicate that calculators have become ubiquitous among middle school students and their families, and may perhaps be regarded as commonplace items, like notebooks or pens.

While positive results were found, further research is needed to determine the most effective ways to integrate calculators into the mathematics curriculum. Research is needed to determine how calculators improve mathematics performance; is it merely computational efficiency, or are more subtle cognitive and affective factors involved? Some sort of protocol analyses might be appropriate. Parts of this study should be replicated using assessment instruments with items designed to avoid calculator bias, and on standardized tests designed for calculator use.

There should be further research regarding the attitudes and beliefs of mathematics students. This study's findings suggest a connection between female students' feelings of self-confidence in mathematics and calculator use. The issue of students' confidence and feelings of self-esteem in mathematics as affected by calculators and other empowering technologies warrants continued investigation. Research could be undertaken to analyze appropriate use versus mindless "overuse" of the calculator, and to determine whether students use calculators for situations that would be better approached by other solution techniques, such as estimation or mental calculation.

Specific research questions about calculator pedagogy are only beginning to find answers. As new calculator functions emerge, such as fraction keys, research can test their effectiveness and pedagogical soundness. Fundamental questions, as well, remain unanswered. For instance, further research is needed to ascertain the best uses of the calculator in the curriculum. Are there prerequisite skills students should have before they begin using a calculator? Project results raise questions about how to effectively train and motivate teachers to integrate calculators. Specific staff development research results could help overcome the inertia of some teachers when faced with new technologies.

The NCTM vision of integrated mathematics education will take longer to realize if we persist in the fear that calculators will prevent students from learning basic arithmetic. The results of this study suggest that calculators have become commonplace among middle school students, and that students perform as well if not better, even on a standardized test, using a calculator. If, as the National Research Council (1989) suggests, students who use calculators emerge from school with "better problem solving skills and much better attitudes about mathematics" (p. 48), we must continue to explore and promote the improvement of mathematics education through the use of a calculator-enhanced curriculum.

References

Bitter, G. G. (1980). Calculator teacher attitudes improved through inservice education. *School Science and Mathematics, 80*(4), 323-326.

Cronbach, L. J. (1984). *Essentials of psychological testing.* New York: Harper & Row.

Dick, T. (1988). The continuing calculator controversy. *Arithmetic Teacher, 35*(8), 37-41.

Dutton, W. H., & Adams, L. S. (1961). *Arithmetic for teachers.* Englewood Cliffs, NJ: Prentice-Hall.

Hart, L., & Stanic, G. (1989, April). *Attitudes and achievement-related behaviors of middle school mathematics students: Views through four lenses.* Paper presented at the annual meeting of the American Educational Research Association, San Francisco, CA.

Hembree, R., & Dessart, D. (1986). Effects of hand-held calculators in precollege mathematics education: A meta-analysis. *Journal for Research in Mathematics Education, 17,* 83-99.

Hope, J., & Sherrill, J. (1987). Characteristics of unskilled and skilled mental calculators. *Journal for Research in Mathematics Education, 18,* 98-111.

Lappan, G., Reyes, L., & Stanic, G. (1988). Gender and race equity in primary and middle school mathematics classrooms. *Arithmetic Teacher, 35*(8), 46-48.

National Council of Teachers of Mathematics. (1989). *Curriculum and evaluation standards for school mathematics.* Reston, VA: Author.

National Research Council. (1989). *Everybody counts: A report to the nation on the future of mathematics education.* Washington, DC: National Academy Press.

Suydam, M. (1982). Update on research on problem solving: Implications for classroom teaching. *Arithmetic Teacher, 29*(6), 56-60.

Suydam, M. (1990). *Research on the use of calculators in mathematics instruction.* Unpublished paper.

Szetela, W., & Super, D. (1987). Calculators and instruction in problem solving in grade 7. *Journal for Research in Mathematics Education, 18,* 215-229.

Research on Calculator Use in Middle School Mathematics Classrooms

Juanita V. Copley
Susan E. Williams
Shwu-Yong L. Huang
Hersholt C. Waxman
University of Houston

The use of calculator technology has been a frequent focus in the research area of mathematics instruction (Mathematical Sciences Education Board, 1989, 1990, 1991; Szetela & Super, 1987; Suydam, 1982; Williams, Copley, Huang, & Bright, 1993; Willoughby, 1990). In a recent position statement, the National Council of Teachers of Mathematics (NCTM) recommended that "publishers, authors, and test writers integrate the use of calculators into their mathematics materials at all levels" (NCTM, 1991, p. 8). Hembree and Dessart's (1986) meta-analysis and Bitter and Hatfield's (1993) study found that the use of calculators produced higher achievement scores both in basic operations and in problem solving. In addition, the researchers concluded that students using calculators possess a better attitude toward mathematics than non-calculator students. Hembree and Dessart concluded their study by stating that "it no longer seems a question of *whether* calculators should be used along with basic skills instruction, but *how*" (Hembree & Dessart, 1986, p. 87).

Although the use of calculators in mathematics classrooms has been widely advocated by educators and professional organizations, there have been few research studies that have specifically examined how mathematics teachers use calculators in their classes. In most studies, the effects of calculator usage are described using only self-reported survey instruments, student outcome measures, and attitudinal instruments. Few, if any, studies report observational data of calculator usage in mathematics classrooms. Proponents of integrating calculator technology into the mathematics curriculum suggest that integration is more than just a physical change that can be described by counting the number of calculators purchased by a school district or by reporting the number of calculator inservice sessions available to teachers (Hill, 1980).

Process-product studies which investigated linkages between specific teaching behaviors in classrooms and student learning outcomes have been perceived as beneficial to research in mathematics education (Romberg & Carpenter, 1986). To focus on the specific instructional strategies that positively affect student achievement, the mere study of outcomes has not been sufficient. A focus on the classroom processes that mediate student outcomes has been proposed as essential to research involving mathematics education (Shavelson, Webb, Stasz, & McArthur, 1988). Good and Biddle (1988) contend that if intervention programs are observed as they are implemented in classrooms, a composite of variables and their interactions can be analyzed.

Research indicates that the implementation of calculator technology in the mathematics curriculum is dependent on several factors. Teachers' attitudes toward the use of specific technology is positively correlated to the implementation of technology (Office of Technology Assessment, 1988; Terranova, 1990). Bitter and Hatfield (1993) found that teachers' commitment to implementing calculators in mathematics instruction affected students' performance on all ITBS mathematics subtests. Several studies have examined curriculum development and implementation and found that innovations may result in disappointing outcomes, not because of inadequacies of the innovative idea, but because of the lack of teacher involvement in the development of the innovation (Berman & McLaughlin, 1976; Charters & Pellegrin, 1976; Stein & Wang, 1988; Strathe & Hatcher, 1986).

Project Objectives

The series of studies presented in this paper were conducted during the first year of a U.S. Department of Education, Dwight D. Eisenhower Mathematics and Science Education Program grant which investigated calculator use and calculator curriculum development in middle school mathematics classrooms. The purposes of these studies were to (a) describe the quantity of calculator use, the types of activities with which calculators were used, and how teachers used calculators for instructional purposes, (b) investigate teachers' attitudes toward calculators and examine whether teacher characteristics effect their attitudes, (c) determine the effect, if any, of calculator use and instruction on student mathematics problem-solving achievement, and (d) investigate the effect of teacher involvement in the curriculum development of calculator materials on the implementation of calculators in the classroom.

Methods

Sample

The subjects in this series of studies were 56 middle school mathematics teachers (i.e., grades 6, 7, & 8) and nearly 6,800 of their students from a large, multi-ethnic, metropolitan school district located in a large city in the south central region of the United States. The teachers had taught for an average of 11.2 years, ranging from 0 (beginning teachers) to 26 years. Slightly more than 80% of the teachers were female and nearly 20% of them were male. About 55% of the teachers had completed a master's or higher degree, 45% of them had obtained a bachelor's degree. Nearly 40% of the teachers were teaching 6th grade, 30% were teaching 7th grade, and over 30% 8th grade. All of the teachers received a minimum of 12 hours of in-service training on the use of calculators during the fall semester. Each of the 6,800 middle school students in the district was issued a calculator at the beginning of the spring semester.

From the original group of 56 teachers, 22 teachers were selected by the school district to be curriculum writers. These teachers were released from their classroom teaching assignments a half day, once a month for eight months during the school year. With mathematics educators from a nearby College of Education, the writers developed and field-tested a set of mathematical explorations that required the use of a calculator. The developed explorations were correlated with the existing curriculum and the *Curriculum and Evaluation Standards for School Mathematics* (NCTM, 1989).

Instruments

For this series of studies, one instrument was used to observe and assess calculator instruction in mathematics classrooms, two instruments were administered at the student level to measure problem-solving achievement before and after the first year, and one instrument was administered to individual teachers to measure teacher attitude toward calculators.

Observation instrument. An Observation Rating Scale for Calculator Implementation (Williams, Waxman, & Copley, 1990) was used to describe and measure the amount of calculator instruction in each classroom and to assess the quality of that instruction. Seventeen indicators measure the quantity and quality of calculator instruction and use. The inter-rater reliability for the instrument was greater than 0.80.

Preachievement measure. The Essential Elements of Elementary Mathematics Test (White, 1986) was administered to all middle school students as a preachievement measure and used as a covariate in one of the analyses. The test was developed to assess mathematics achievement in the Texas State Board of Education Essential Elements for Grades K-8 (1984). The test requires little computation and assesses a higher level of learning than recognition or recall of basic algorithms. The test consists of 50 multiple-choice paper-and-pencil items. It was designed to sample the following seven categories of the Essential Elements:
1. concepts and skills associated with the understanding of numbers and the place-value system (12 items)
2. basic number operations, their properties, and their uses (13 items)
3. solving problems by selecting and matching strategies to given situations (9 items)
4. measurement concepts and skills using metric and customary units (7 items)
5. properties and relationships of geometric shapes and their applications (2 items)
6. representation of numbers on a line and pairs of numbers on a coordinate plane (2 items)
7. use of probability and statistics to collect and interpret data (5 items)

A sixth grade mathematics textbook was used to check the appropriate level of the test items. The test was administered to 338 preservice elementary education teachers. For that sample, the reliability estimate using the Kuder-Richardson Formula 20 was 0.87 (White, 1986).

Four-Step Problem-Solving Test. The Four-Step Problem-Solving Test (Hofmann, 1986) was administered to middle-school students as a post achievement test of problem-solving skills. The multiple-choice test consists of 10 nonroutine problems each with four related questions. The design of the test is based on a four-step heuristic: (a) read to understand the problem, (b) select a strategy, (c) solve, and (d) review and extend. The first question (Subscale 1) for each problem refers to information necessary for determining the solution. The second question (Subscale 2) asks for a strategy to solve the problem. The third question (Subscale 3) offers a possible solution. The fourth question (Subscale 4) refers to an extension of the original problem or asks a related question based upon facts stated or implied by the original problem. The reliability of the total test for sample sizes larger than 19 was reported as ranging from .69 to .85 (Hofmann, 1986).

Teacher attitude scale. A revised version of the Pocket Calculator Attitude Scale (Bitter, 1980) was used to assess teachers' attitudes toward calculators in general and toward students' calculator use for learning mathematics. All middle school teachers responded to the scale consisting of 20 statements and indicated their level of agreement with each statement by answering either SD (strongly disagree), D (disagree), NS (not sure), A (agree), or SA (strongly agree). The scale was previously found to be reliable (r=.95) and valid.

Procedures

The Calculator Attitude Scale was administered to all middle school mathematics teachers at the beginning of the fall semester. A three-way multivariate analysis of variance was used to determine if there were any statistically significant differences (p<.05) on attitudes: (a) among teachers of the three grade levels, (b) among teachers with different experience levels, and (c) between two educational levels of teachers.

All teachers were observed at four separate times throughout the school year during their 45-minute mathematics class period. Teachers were unaware that they would be observed during the randomly-selected period. Observers rated each item using a five point scale to indicate the amount of time each behavior occurred. A high score for an item or indicator revealed that teachers were frequently observed using the particular behaviors. Students were administered the Essential Elements Test at the beginning of the school year; the Problem-Solving Test was administered at the end of the school year.

Mean scores for the calculator instructional indicators listed on the Observation Rating Scale were derived for the final two observations along with the means for the Essential Elements Test and the Problem-Solving Test by teacher. Setwise multiple regression analyses were then used to explain how the indicators affected student problem-solving achievement after controlling for initial achievement.

To investigate the implementation process of writers (N=22) and non-writers (N=22), teachers were randomly chosen from the non-writing middle school mathematics teachers in the district. Means and standard deviations for the indicators of the observation instrument for both the experimental (writers) and control (non-writers) groups were derived, and *t*-tests were used to compare the implementation of calculators of the two groups on each of the 17 indicators.

Results of the Descriptive Studies

Although all middle-school students were individually issued a calculator, there was no guarantee that the students would bring the calculators to class each day nor that the teachers would provide opportunities for the students to use them. Thirty percent of the 224 observations reported that there were no calculators at the desks or tables of students in the classroom. Of the remaining 157 observations, the overall findings indicate that a mean of 61% of the students had calculators at their desks or tables. Of those students that had calculators, calculators were used an average of 43% of the time.

Using only the observations which indicated that calculators were used (N=157), the results reveal that calculators were primarily used for computation-focused activities (66%) and for verifying answers (28%). Two particular activities, exploration and the solution of non-routine problems, were used by teachers and students on a very limited basis (6%) during the classroom observations. The results are listed in Table 1.

Table 1. *Means and Standard Deviations of the Percent of Time Behavior was Observed as Described by Indicators for Types of Calculator Activities (N=157)*

Observed Behaviors	M	SD
Students use calculators for computation.	65.93	40.15
Students use calculators for exploration/induction.	5.73	18.53
Students use calculators for solving routine word problems.	14.49	29.71
Students use calculators for solving non-routine problems.	5.73	20.39
Students use calculators for self-checking/verifying answers.	28.34	39.64
Students use calculators for games.	1.91	12.19

The descriptive data also indicate pertinent information regarding the quality of instruction when teachers use calculators. Generally, teachers initiated the use of the calculator approximately 33% of the time and allowed students to determine the appropriate use of the calculator about 40% of the time. The smallest percentage, 7%, was received for "emphasizes importance of estimation to determine reasonableness of calculator answer." Table 2 lists the other indicators as well as the means and standard deviations. It is important to note that for most indicators the standard deviations are quite large, suggesting that there was a great deal of variance in the way calculators were used by the teachers.

Table 2. *Means and Standard Deviations of the Percent of Time Behavior was Observed as Described by Indicators for Teacher Calculator Instruction (N=157)*

Observed Behaviors	M	SD
Allows students to determine appropriate use of calculator.	39.65	43.12
Emphasizes importance of estimation to determine reasonableness of calculator answer.	7.32	21.97
Explains relationship between calculator and paper-and-pencil algorithm.	10.82	25.52
Stresses use of calculator as time-saver.	15.13	30.59
Stresses use of calculator as a problem-solving tool.	11.31	24.91
Teacher demonstrates use of calculator.	15.13	30.59
Students use calculators during teacher demonstration.	18.47	30.38
Student initiated use of calculator.	32.64	38.88
Teacher initiated use of calculator.	33.44	39.18

Table 3 reports the teachers' responses on the Calculator Attitude Scale. Overall, the findings revealed that the teachers were generally positive toward calculator use. However, less than half of the teachers indicated that students should be allowed to use calculators while taking mathematics tests. Approximately 20% of the teachers expressed concern about students' use of calculators because of their lack of understanding of basic computation skills or their ability to do mental computations.

Table 3. *Teachers' Responses on the Calculator Attitude Scale in Percentages*

Item	SD/D	NS	A/SA
Calculators should be an integral part of the curriculum.	0	14	86
I get no satisfaction from using calculators.	91	7	2
I want calculators for all students.	0	11	89
Calculators are too expensive for classroom use.	82	11	7
Calculators are neat.	2	7	91
The use of calculators for games and fun should be encouraged.	5	2	93
Students should not be allowed to use calculators while taking math tests.	57	20	23
I have a growing appreciation of calculators through understanding their application to the school curriculum.	5	11	84
I have never liked calculators.	96	2	2
Working with calculators is fun.	0	9	91
I am afraid to work with calculators and use them in class with students.	82	11	7
Calculators will cause students to not understand the basic computation skills.	58	23	19
Calculators should be available for all students in all grades.	12	12	76
I don't feel calculators should be allowed in the schools	96	2	2
Working with calculators is boring.	93	7	0
The use of calculators is causing students to lose the chance to do mental computations in school.	58	21	21
Calculators can stimulate a child to study mathematics.	5	16	79
Calculators do not allow students to do simple mathematics on paper.	62	12	26
Calculators make mathematics fun.	5	7	88

Notes: SD/D=Strongly Disagree/Disagree; NS=Not Sure; A/SA=Agree/Strongly Agree

Results of Comparative-Correlational Studies

To compare the attitudes of teachers with different characteristics, multivariate analyses of variance (MANOVA) and analyses of variance (ANOVA) were used. The MANOVA results indicated that while there were no overall significant differences, there were significant differences ($p<.05$) on some individual attitude items by grade level taught and teachers' educational level. The ANOVA results by grade level taught revealed that eighth grade teachers were significantly stronger ($F=3.27$, $p<.05$) than those of 6th and 7th grade teachers on the attitude item, "Calculators can stimulate a child to study mathematics." In addition, the mean of the 8th grade teachers was significantly stronger ($F=4.25$, $p<.05$) than that of the 6th grade teachers on the attitude item, "Calculators should be an integral part of the curriculum." Also, teachers with master's degrees had significantly more positive attitudes toward calculator usage in school than did teacher's with bachelor's degrees.

Table 4 reports the results of the setwise regression of problem-solving achievement on preachievement and types of calculator activities. As anticipated, preachievement has the largest significant effect (Beta=.81; p<.001) on total achievement as well as on the subscales (Beta=.78, .75, .81, .81 respectively; p<.001). The activity, "use calculators for self-checking/verifying answers" exhibited a nearly-significant, positive effect on understanding the problem (SS1), a strongly significant, positive effect on correctness (SS3), and a significant positive effects on extension (SS4) as well as on the total problem-solving score (Total). The activity, "use calculators for solving routine word problems" had a nearly significant, negative effect on correctness (SS3) and extension (SS4). The activity, "use calculators for exploration/induction" had a nearly significant, positive effect on extension (SS4).

Table 4. *Regression of Problem-Solving Achievement on Preachievement and Types of Calculator Activities*

	Beta values				
Independent Variables	SS1	SS2	SS3	SS4	Total
Preachievement	.78***	.75***	.81***	.81***	.81***
Students use calculators for exploration/induction.	.08	.07	.11	.14@	.11
Students use calculators for solving routine word problems	-.12	-.05	-.13@	-.12@	-.11
Students use calculators for solving non-routine problems.	.01	.06	.04	.02	.03
Students use calculators for self-checking/verifying answers.	.17@	.10	.20**	.17*	.17*
R^2 values	.73***	.63***	.78***	.81***	.79***

@ p<.10, *p<.05, **p<.01, ***p<.001
Notes: SS1 = understanding the problem, SS2 = strategies to solve problem, SS3 = correctness of answer, SS4 = extension of answer

Table 5 reports the results of the stepwise regression of problem-solving achievement on preachievement and calculator instructional indicators. Other than the expected large significant effect of Preachievement, only one instructional indicator had a significant effect on problem-solving achievement. Instruction which stresses the use of the calculator as a "time-saver" had a significant, negative effect on extension (SS4). There were no other nearly-significant or significant results for this regression.

Results of the Quasi-Experimental Study

Overall, eight of the 17 findings revealed significant differences between the experimental and control groups on calculators use in the classroom (Table 6). Experimental teachers were observed significantly more often than control teachers to (a) allow students to determine appropriate use of the calculator, (b) emphasize the importance of estimation for determining the reasonableness of a calculator answer, and (c) stress the use of the calculator as a problem-solving tool. Students in the classrooms of the experimental group teachers used calculators significantly more often than their counterparts in the control classrooms for the following purposes: (a) computation, (b) exploration and induction, (c) solving routine word problems, and (d) verifying answers. Although all students in the district were individually issued a calculator, there was a significant difference between the percent of students who brought their calculators to class. Only about 47% of the students from the control teachers' classrooms were observed having calculators at their desks, while 60% of the students in experimental teachers' classrooms were observed having them.

Table 5. *Regression of Problem-Solving Achievement on Preachievement and Calculator Instructional Indicators*

Independent Variables	Beta values				
	SS1	SS2	SS3	SS4	Total
Preachievement	.78***	.79***	.81***	.81***	.82***
Allows students to determine appropriate use of calculator.	.12	-.05	.10	.09	.07
Emphasizes importance of estimation to determine reasonableness of calculator answer.	-.06	.01	-.02	.00	-.02
Explains relationship between calculator and paper and pencil algorithm.	.02	.12	.03	.00	-.04
Stresses use of calculator as a "time-saver".	-.07	.04	-.13	-.19*	-.10
Stresses use of calculator as a "problem-solving tool".	.08	.05	.07	.08	.07
Student initiated use of calculator.	-.04	-.01	.00	-.02	-.02
Teacher intiated use of calculator.	-.07	-.07	.02	.06	-.01
R^2 values	.71***	.63***	.78***	.79***	.76***

*p<.05,   ***p<.001
Notes: SS1 = understanding the problem, SS2 = strategies to solve problem, SS3 = correctness of answer, SS4 = extension of answer

Discussion

The results of this series of studies suggest implications for further education of mathematics teachers. Even though teachers and students have calculators issued to them, they are not always used or even available for use. Perhaps this is an indication that calculators are only appropriate for some mathematics activities or perhaps teachers and students are not aware of the variety of possible uses for calculators. The solution of non-routine problems and exploratory activities are both activities that are recommended by experts as a way to develop mathematical meanings (Szetela & Super, 1987). Calculators are uniquely suitable for these types of activities. In fact, the positive effect of exploratory activities on achievement in problem-solving extensions along with the negative effects of routine problem-solving activities indicate that exploratory activities and the use of non-routine problems are activities that have potential. The low percentage of time that these activities were observed indicate that teachers need to be trained to recognize the possibilities when calculators are added as tools to learn mathematics.

Table 6. *Comparison of Experimental and Control Groups of Middle School Mathematics Teachers on the Implementation of Calculators in the Classroom*

Observed Behavior	Experimental (N=22)		Control (N=22)		t
	M	SD	M	SD	
Percentage of students who have calculators at their desks/tables.	59.09	14.77	47.27	21.64	2.12*
Amount of time students (who have calculators) use them.	43.64	12.32	36.36	16.77	1.55
Students use calculators for computation.	63.64	20.01	44.32	28.80	2.58*
Students use calculators for exploration/induction.	14.77	16.65	4.55	9.87	2.48*
Students use calculators for solving routine word problems.	21.59	19.36	10.23	16.65	2.09*
Students use calculators for solving non-routine problems.	9.09	12.31	4.55	9.87	1.35
Students use calculators for self-checking/verifying answer.	36.36	16.77	22.73	18.75	2.54*
Students use calculators for games.	4.55	9.87	1.14	5.33	1.43
Allows students to determine appropriate use of calculator.	45.45	19.88	27.27	24.29	2.72*
Emphasizes importance of estimation to determine reasonableness of answer.	14.77	19.91	4.55	9.87	2.16*
Explains relationship between calculator and paper-and-pencil algorithm.	19.32	17.13	11.36	12.74	1.75
Stresses use of calculator as "time-saver".	21.59	15.99	12.50	14.94	1.95
Stresses use of calculator as a problem-solving tool.	17.05	16.16	6.82	11.40	2.43*
Teacher demonstrates use of calculator.	21.59	22.22	15.91	18.17	0.93
Students use calculators during teacher demonstration.	23.86	24.97	14.77	16.65	1.42
Student initiated use of calculator.	35.23	16.65	27.27	18.75	1.49
Teacher initiated use of calculator.	38.64	22.79	26.14	23.75	1.78

*p<.05

One particular indicator, "the emphasis of the importance of estimation to determine reasonableness of answers," is an accepted behavior considered to be essential to effective calculator instruction (Hill, 1980). The finding that in over 90%

of the classrooms, the selection "not observed" was made for that indicator is troublesome. Estimation as a method to determine the reasonableness of an answer needs to be a focus in teacher training programs. In addition, the significant, positive effects of using calculators to verify answers on problem-solving achievement imply that teachers should teach students to analyze their answers for both reasonableness and reliability.

One of the most positive results of this study are the teachers' general attitudes toward calculator use. It is encouraging to note that 84% of the teachers felt a growing appreciation of calculators through understanding their applications to the school program. The finding that years of teaching experience made no significant difference on teachers' attitudes toward calculators but that teachers with graduate degrees had significantly more positive attitudes implies that graduate-level teacher education may provide information and experiences that positively affect teachers' attitudes toward change and, in particular, the use of calculators. Because of the direct relationship of attitudes to their implementation of the use of calculators in mathematics classes and to the improvement of students' mathematics achievement (Bitter & Hatfield, 1993; Mathematical Sciences Education Board, 1989), it is important that all education decision-makers cultivate an environment which encourages mathematics teachers to pursue higher degrees.

Finally, the results of the study of non-writers and writers suggest that the presence of technology and quick treatments (12 hours of calculator inservice) do not necessarily insure immediate changes in instructional processes. The results do indicate, however, that teacher involvement in the change process (even if that involvement is minimal) seems to have a significant effect on the implementation of calculators.

Student use of the calculators may be the most important issue for research purposes. In this study, the predominant use of calculators from the six available indicators was computation ($C=44\%$, $E = 64\%$). More specific information about the computational use of calculators, however, is required before a judgment regarding the quality of calculator use in these classrooms can be made. In retrospect, the purpose of the computation being performed by the calculator is actually more important than the fact that on average calculators were used for computation approximately 64% of the time in experimental classrooms and 44% of the time in control classrooms.

The findings from this study suggest that the compressed two-day in-service training in the fall did little to affect the implementation process. Teacher involvement in the writing of technology-based curriculum seems to have had a much greater impact. The writers used the innovation more, and also used it in more effective ways. Future studies involving long-range implementation of developed curriculum by writers and non-writers, as well as other aspects of the implementation process involving technology, require investigation. Additional calculator indicators and quality constructs for effective calculator instruction need to be developed to provide a more complete description of calculator use in classrooms.

Acknowledgment

Support for this research was provided in part by a grant from the U.S. Department of Education, Dwight D. Eisenhower Mathematics and Science Education Program, award number R168D00311. Opinions and views expressed in this paper do not necessarily reflect those of the granting agency.

References

Berman, P., & McLaughlin, M. (1976). Implementation of educational innovation. *Educational Forum, 40,* 345-370.

Bitter, G. G. (1980). Calculator teacher attitudes improved through inservice education. *School Science and Mathematics, 80,* 323-326.

Bitter, G. G., & Hatfield, M. M. (1993). Integration of the Math Explorer calculator into the mathematics curriculum: The calculator project report. *Journal of Computers in Mathematics and Science Teaching, 12*(1), 59-81.

Charters, W., & Pellegrin, R. (1976). Barriers to the innovation process: Four case studies of differentiated staffing. *Educational Administration Quarterly, 9,* 3-25.

Good, T. L., & Biddle, B. J. (1988). Research and the improvement of mathematics instruction: The need for observational resources. In D. Grouws, T. Cooney, & D. Jones (Eds.), *Perspectives on research on effective mathematics teaching* (pp. 114-142). Reston, VA: National Council of Teachers of Mathematics.

Hembree, R., & Dessart, D. J. (1986). Effects of hand-held calculators in precollege mathematics education: A meta-analysis. *Journal for Research in Mathematics Education, 17,* 83-89.

Hill, S. A. (1980). Recommendations for school mathematics programs of the 1980s. In M. M. Lindquist (Ed.), *Selected issues in mathematics education* (pp. 258-268). Washington, DC: National Education Agency.

Hofmann, P. S. (1986). *Construction and validation of a testing instrument to measure problem-solving skills of students.* Unpublished doctoral dissertation, Temple University, Philadelphia, PA.

Mathematical Sciences Education Board. (1989). *A challenge of numbers: People in the mathematical sciences.* Washington, DC: National Academy Press.

Mathematical Sciences Education Board. (1990). *Reshaping school mathematics: A philosophy and framework for curriculum.* Washington, DC: National Academy Press.

Mathematical Sciences Education Board. (1991). *Counting on you: Actions supporting mathematics teaching standards.* Washington, DC: National Academy Press.

National Council of Teachers of Mathematics (1991). *Professional standards for teaching mathematics.* Reston, VA: Author.

Office of Technology Assessment. (1988). *Power on! New tools for teaching and learning.* Washington, DC: U.S. Government Printing Office.

Romberg, T. A., & Carpenter, T. P. (1986). Research on teaching and learning mathematics: Two disciplines of scientific inquiry. In M. C. Wittrock (Ed.), *Handbook of research on teaching* (3rd ed., pp. 850-873). New York: Macmillan.

Shavelson, R., Webb, N., Stasz, C., & McArthur, D. (1989). Teaching mathematical problem solving: Insights from teachers and tutors. In R. Charles

& E. Silver (Eds.) *The teaching and assessing of mathematical problem solving* (pp. 203-231). Reston, VA: National Council of Teachers of Mathematics.

Stein, M. K., & Wang, M. C. (1988). Teacher development and school improvement: The process of teacher change. *Teaching & Teacher Education, 4*, 171-187.

Strathe, M. I., & Hatcher, C. W. (1986). Curriculum development by consensus: An evaluation study of model implementation. *Planning and Changing, 17*(2), 79-90.

Suydam, M. N. (1982). *The use of calculators in pre-college education: Fifth annual state-of-the-art review.* Columbus, OH: Calculator Information Center. (ERIC Document Reproduction Service No. ED 220 273)

Szetela, W., & Super, D. (1987). Calculators and instruction in problem solving in grade 7. *Journal for Research in Mathematics Education, 18*, 215-229.

Terranova, M. E. (1990, May). *Elementary teachers' and principals' feelings and beliefs about calculator use.* Paper presented at the Annual Meeting of the New England Educational Research Organization, Rockport, ME. (ERIC Document Reproduction Service No. ED 325 371)

White, M. A. (1986). *Preservice teachers' achievement in the Essential Elements of elementary school mathematics: The development of an evaluation instrument.* Unpublished doctoral dissertation, University of Houston, Houston, TX.

Williams, S. E., Copley, J. C., Huang, S. L., & Bright, G. W. (1993). Effect of teacher involvement in curriculum development on the implementation of calculators. *Journal of Technology and Teacher Education, 1*(1), 53-62.

Williams, S. E., Waxman, H. C., & Copley, J. R. (1990). *The Observation Rating Scale for Calculator Implementation.* Houston, TX: University of Houston.

Willoughby, S. S. (1990). *Mathematics education for a changing world.* Alexandria, VA: Association for Supervision and Curriculum Development.

The Graphing Calculator in Pre-Algebra Courses: Research and Practice

Paul A. Kennedy
Southwest Texas State University

The graphing calculator is a powerful tool for problem solving and for developing algebraic concepts. It can be an integral part of discovery teaching, an inductive/deductive process that maximizes student learning. This inductive/deductive approach to teaching involves the teacher as facilitator asking the right questions and providing specific examples, so that students can make their own generalizations through discovery and hence develop the concepts necessary to make appropriate applications. Graphing calculator use in pre-algebra courses allows students to access algebraic concepts from an intuitive perspective. "Weakness in ... skills need no longer prevent students from understanding ideas in more advanced mathematics." Using technology enables students to "explore mathematics on their own, to ask countless 'what if' questions" (National Research Council, 1989, p. 62).

This article highlights graphing calculator use in Partnership for Access to Higher Mathematics (PATH Mathematics), a project funded by the U.S. Department of Education. In particular, this paper focuses on the use of concrete and real-world examples that can be represented by patterns, tables, and graphs and on some preliminary research findings from the project.

PATH Mathematics, a program for at-risk off-track eighth and ninth graders, was designed to provide them access to Algebra I and beyond. Pre-algebra students include all middle school and off-track high school students who have not reached Algebra I. All of the activities presented have been successfully field-tested with pre-algebra students and/or teachers of pre-algebra students. Cooperative learning is a major component of the curriculum. Students typically work in pairs within teams of four in an activities-based curriculum.

The PATH curriculum is being developed using digital fiber optics interactive television, collaboratively between the university and the mathematics department of a local school district. The calculator intensive curriculum centers on the appropriate use of manipulatives and technology, coupled with a problem-solving focus and cooperative learning to enable students to demonstrate (a) proficiency in quantitative reasoning, (b) proficiency in proportional reasoning, (c) conceptual understanding of algebraic concepts and operations, (d) conceptual understanding of geometry, (e) ability to solve measurement problems, (f) ability to solve elementary statistical and probability problems, and (f) ability to solve problems involving patterns, relations and functions. The graphing calculator is used throughout the curriculum as a tool for computation, concept development, and problem solving.

Related Research

Since its recent arrival in the mathematics classroom, the graphing calculator has been largely used in the advanced mathematics courses. Most studies have focused on the use of these calculators in college and high school pre-calculus courses. The results tend to show no significant gains in achievement through calculator use, but improvement in attitude, better understanding of geometric representations of algebraic problems, and a stronger inclination toward exploration, conjecturing, and generalizing have been noted (Army, 1991; Becker, 1991; Giamati, 1990; Rich, 1989).

Can the graphing calculator be used to enhance pre-algebra students' algebraic thinking over traditional methods through the integration of the calculator as a tool to facilitate concept development and problem solving? Embse (1992) suggests ways to approach pre-algebra concepts, particularly the variable concept, with the graphing calculator to "explore patterns and processes and to solve problems" (p. 65). In a recent graphing calculator study of linear functions with eighth graders, Vazquez (1990) noted no significant difference in achievement but significant gains in attitude and spatial visualization skills by the treatment group.

In PATH Mathematics we are primarily concerned with access. Should we even use the graphing calculator in pre-algebra, particularly with off-track students? Or, put another way, should those students who have been traditionally tracked out of mathematics be offered the same access to technology and higher mathematics that we offer students in advanced classes? Can appropriate calculator use make a difference in providing that access for all students? Is the graphing calculator a tool for more effectively developing algebraic concepts?

The Electronic Chalkboard

The electronic chalkboard, or multiline display feature of the graphing calculator facilitates concept development and problem solving by making it possible to examine a mathematical situation sequentially, since as many as four expressions with corresponding values can be viewed at the same time. Because expressions are entered in the way that students would ordinarily write them and can be rewritten for editing with a keystroke, students can evaluate expressions in problem situations quickly while reinforcing algebraic structure. Suppose, after some work with manipulatives, we are investigating products of integers by looking at the patterns developed in Figure 1. The multiline feature of the electronic chalkboard allows students to use the patterns to generalize the rules for multiplication. In addition students can see how to represent multiplication using parenthetical notation. This is important later when we develop the notion of variable using tables and patterns.

Technology provides ways for students to inductively investigate mathematical concepts that would otherwise not be possible. Using the electronic chalkboard feature of the graphing calculator, the teacher can create sequences of expressions that allow students to identify patterns and generalize rules. Since the calculator follows order of operations, a sequence of expressions can be evaluated with the calculator, thus enabling students to generalize the rules for order of operations: 2+3*4; 2-3*5; 4*2-3*4; 3+2*5-4; 6/2*3; 14/2+3*3-2; 2-3(5-4)-6*2; 3*2^4. Students can then compile a rule that agrees with the timeless "Please Excuse My Dear Aunt Sally."

```
| 2(2)  |    |
|       |  4 |
| 2(1)  |    |
|-------|----|
|       |  2 |
| 2(0)  |    |
|       |  0 |
| 2(-1) |    |
|       | -2 |
| 2(-2) |    |
|       | -4 |
```

```
| 2(3)  |    |
|       |  6 |
| 1(3)  |    |
|-------|----|
|       |  3 |
| 0(3)  |    |
|       |  0 |
| -1(3) |    |
|       | -3 |
| -2(3) |    |
|       | -6 |
```

```
| -2(2)  |    |
|        | -4 |
| -2(1)  |    |
|--------|----|
|        | -2 |
| -2(0)  |    |
|        |  0 |
| -2(-1) |    |
|        |  2 |
| -2(-2) |    |
|        |  4 |
```

Figure 1. *Multiplication patterns*

In another sequence, students might investigate the notion of principle square root. Suppose after developing the concept of area concretely, students are posed with the problem of a contractor being asked to pour a patio slab that has area 10 square meters. Students are aware that a 3x3 square has area 9 and a 4x4 square has area 16. So the side length lies between 3 and 4. Using a calculator and guess and check students explore the problem using an iterative approach (e.g., Nichols, Litwiller, & Kennedy 1992, p. 450). Exploring further, students find that because of rounding, several different values can be squared to obtain 10, for example

3.1622776605 and 3.1622776604. Students are now close to the defining characteristic of irrational number. Finally the teacher introduces the square root symbol. This activity reinforces place value, while students develop the concepts of principle square root and irrational number.

The use of table building on the calculator is particularly nice for problem situations that can be represented with variables later. "Suppose a long distance telephone company charges $9.75 per month plus 23.5¢ per minute for long distance service. What would your monthly bill be if you talked for 30 minutes, 60 minutes, 90 minutes, or 120 minutes? Estimate first, then compute. Using compatible numbers, students might first estimate $10 + \frac{1}{4} (32) = 10 + 8 = 18$, which is an overestimate. It is important for students to be able to estimate so that they can accurately interpret calculator results. Figure 2 shows a possible calculator display for this problem.

```
9.75 + .235(30)
---------------------
                 16.8
9.75 + .235(60)
                23.85
9.75 + .235(90)
                 30.9
9.75 + .235(120)
                37.95
```

Figure 2. *Long distance bill*

Later, the numbers in parentheses will be represented by the variable x and the total phone bill by the variable y. This is a good opportunity to pose a few follow-up exploration questions that can be solved using guess and check. Suppose you have budgeted $25 for long distance. How long can you talk? Another long distance company charges 40¢ per minute but no monthly charge. Which company should you use?

In another version of the box problem (Kennedy 1991), students are asked to form open boxes from 8x10 centimeter grids by removing a 1x1, 2x2, and 3x3 square from each corner to form the 1-box, 2-box, and 3-box respectively. They fill the boxes with cubes, generalize the volume formula, compute the volumes of the 1.1 through 1.9 box and finally investigate the maximum volume iteratively with the calculator. Figure 3 shows a typical sequence of displays for the box problem. Students complete the table easily by observing patterns. For example, the lengths decrease by two-tenths: 7.8, 7.6, 7.4, and so on. Students discover that the maximum volume occurs (to the nearest tenth) at the 1.5 box.

The electronic window of the graphing calculator can be used to develop algebraic concepts and to solve problems by building tables of values. Table building and guess and check strategies prepare students to translate numerical expressions into variable expressions, write equations, solve equations and graph functions.

	length	width	height	volume = lwh
1-box	10-2(1)=8	8-2(1)=6	1	48
2-box	10-2(2)=6	8-2(2)=4	2	48
3-box	10-2(3)=4	8-2(3)=2	3	24

Follow the pattern and determine volumes for the 1.1-box 1.2 box, etc.

	length	width	height	volume = lwh
1-box	10-2(1)=8	8-2(1)=6	1	48
1.1 box				
1.2 box				
1.3 box				
1.4 box				
1.5 box				
1.6 box				
1.7 box				
1.8 box				
1.9 box				
2-box	10-2(2)=6	8-2(2)=4	2	48
3-box	10-2(3)=4	8-2(3)=2	3	24

Which box has the most volume? Use your calculator to find the size of the square (to the nearest hundredth) that maximizes the volume.

Figure 3. *Volume table*

Using Concrete Representations to Develop Calculator Skill

The graphing capability can be easily introduced early on by investigating some simple linear relations. Charles (1990), for example, includes a set of pattern problems with cubes (Figure 4). Students are instructed to use cubes to complete the patterns and fill in a table. They discover in the problems that one characteristic changes (variable) and another might remain the same (constant). The idea can be expanded to include plotting the points in the first quadrant, writing a rule where y depends on x, and finally graphing a straight line.

In the first problem students realize that since each story has four cubes and the number of cubes, y, can be obtained by multiplying four times the number of stories, x. Here the graphing features of the calculator are presented to the students. The rule Y = 4X is entered and the RANGE is set from 0 to 95 for X and from 0 to 400 for Y. Divisions of 95 equal units are useful for reading points using TRACE, since the window is 95 pixels wide. Students read the points using TRACE to see if the graph agrees with their own sketch and table. In the second problem the number of cubes in the tower changes while the number of cubes in the base remains the same (3). Here students generalize the rule y = x + 3, sketch a graph and again check with the graphing calculator.

In a number theory exercise, (Fitzgerald, Winter, Lappan, & Phillips, 1986) students are provided with grid paper and wooden squares and instructed to form all rectangles that have area 1, 2, 3, 4, 5, 6, and so on. The dimensions of the rectangle, of course, turn out to be the factors of the number in question. It is interesting to

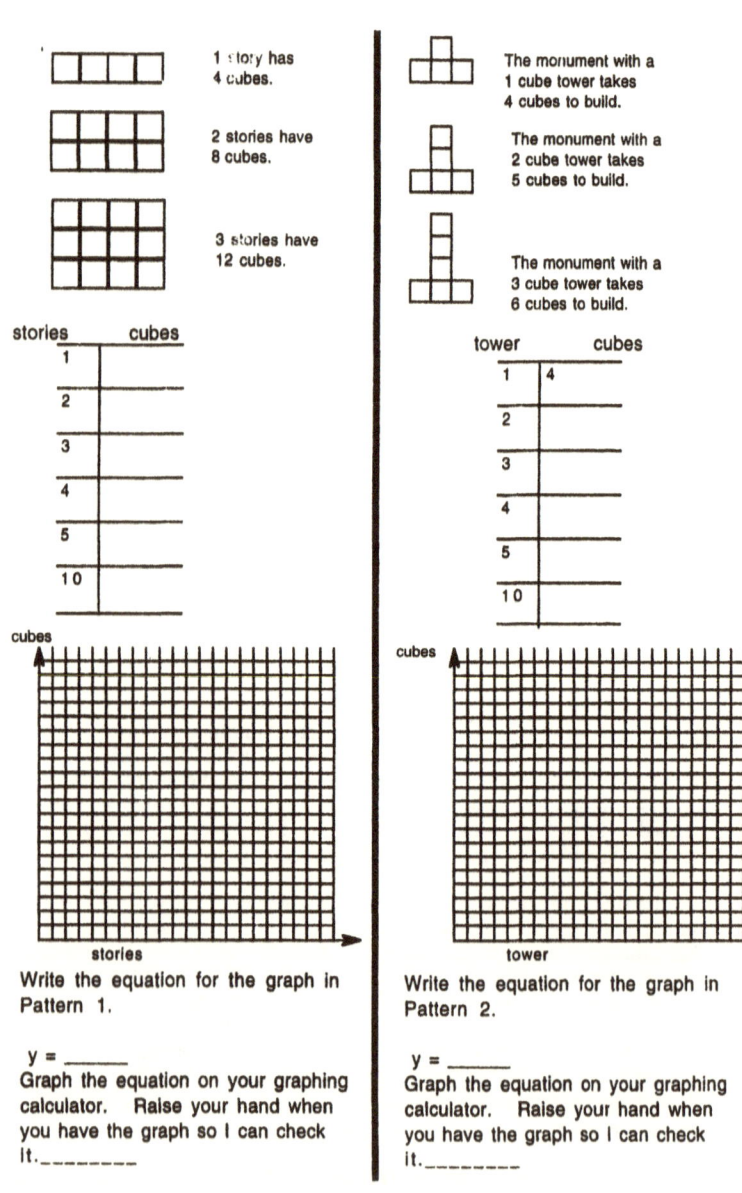

Figure 4. *Concrete models*

extend this problem by using the graphing calculator. For the factors of 18, students would list the factors and then plot the points on the grid and generalize the rule $y = \frac{18}{x}$. When this rule is graphed on the calculator, students discover other non-integer factor pairs. They also see that part of the hyperbole appears in the fourth quadrant. This reinforces the understanding of signed ordered pairs in the plane as well as the rule for multiplying integers.

These simple concrete problems provide excellent opportunities to develop the idea that a variable can represent a quantity that changes and that one variable can depend on another. These activities are important before students get bogged down in solving for x as an unknown, so that they can develop a fuller variable concept. The variable concept is reinforced as students begin to learn how to use a calculator in this simple setting to relate two variables. The graphing calculator is a nice tool for making the link between concrete and graphical representations of problem situations. Students can also easily develop basic calculator skill in this setting - - skill that will become essential for more complex problems.

Patterns to Variables

The multiline feature of the electronic chalkboard allows students to build tables from patterns that can be represented with variables and then graphed. Proportion problems are particularly useful for developing equations of the form y = mx.

Proportional reasoning is critical to success in algebra. As suggested by Post, Behr, and Lesh, (1988) and others, rate problems can be easily understood in terms of using a factor or finding a unit rate. Consider the problems in Figure 5 in which the unit rate is given.

A train averages 40 miles in 1 hour.
How many miles can it travel in 2 hours?
How many miles can it travel in 3 hours?
How many miles can it travel in 15 hours?
How many hours will it take to go 200 miles?
How many hours will it take to go 100 miles?
How many miles can it travel in $4\frac{1}{2}$ hours?

hours	miles
1	40
2	40(2)=80
x	y =

Videos	Cost
x	y =

One video rents for $1.50
How much do 8 videos rent for?
How many videos can be rented for $12?
How many videos can be rented for $526.50?

Figure 5. *Rate problems*

Here students build tables by either multiplying or dividing by the unit rate. Students can easily generalize the rules y = 40x and y = 1.50x, sketch a graph and label the axis, and then graph on the graphing calculator. At some point in this process, students realize there is no longer a need to sketch graphs first. That is, students make a connection between the rule and the graph and can use the calculator to model the problem situation. Students eventually become adept at deciding what RANGE settings are appropriate, but initially students need quite a bit of guidance.

Consider now a problem in which the unit rate is not given.
A salesperson can average 300 miles in 6 hours. At that rate how long will it take her to travel 150 miles? How far can she travel in 5 hours?

The first question can be answered simply by using the factor $\frac{1}{2}$ and multiplying by 6 to obtain 3 hours. The second question is generally solved by computing the unit rate (50 miles in one hour) and then multiplying by the factor 5 to obtain 250 miles. Students easily organize the data in a table and can generalize the rule y = 50x and graph on the graphing calculator. Rate problems provide an excellent vehicle for graphically developing the notion of slope and the equation of the line. Rate problems connect real world experiences to graphical representations to enhance student understanding of how variables relate. To what extent, if any, do graphical representations reinforce the understanding of rate problems?

Consider again the long distance phone problem. From the table of values students can sketch a graph, labeling the x-axis "minutes" and the y-axis "phone bill." It is important for students to connect variables with some real problem at hand. The work done with concrete models (discussed earlier) makes it easy for students to generalize the rule in words (phone bill is $9.75 plus $0.235 times the number of minutes) or in symbols ($Y_1 = 9.75 + \$0.235X$). Writing the rule both in words and in symbols helps students to understand the role of the two variables. Setting the RANGE 0 to 95 for X and 0 to 50 for Y enables the students to check their work. For the second long distance company, without the monthly charge, we have the equation $Y_2 = .40X$. This situation is ideal for higher level teacher questioning. What does the point (0, 25) on the first graph mean? What is the point of intersection? When students graph the two lines together they can TRACE and ZOOM to discover that after 59 minutes it is cheaper to go with the second company. Guess and check, which may become tedious, paves the way for introducing variable expressions for numerical expressions and solving equations that represent the problem situations: 9.75 + .235x = 25 and 9.75 + .235x = .40x. It should be no surprise that students can generalize these rules from real world situations without any formal treatment of linear functions. In the second equation we have essentially solved a system of equations by substitution (without any formalization).

Linear Functions

Real-world examples provide a means for exploring and understanding the Cartesian Plane. With a graphing calculator, students who have been generalizing equations of the form y = mx and y = mx + b from simple concrete and real-world problem situations are ready to explore the parameters "m" and "b" more closely.

Expanding on the video problem from the previous section students can build tables and explore the graphs determined by video cost (y) for videos rented (x) at $1,

$2, $3, and $1.50 each. Slope can be seen simply as the cost per video (rate). Setting RANGE for x from 0 to 95 with XSCL 10 and for y from 0 to 63 with YSCL 10 results in a true-to-scale graph, thus allowing the student to clearly see change in y and change in x. Students develop a sense of slope as steepness and discover easily enough that slope is just the coefficient of x, the unit rate. On the last graph the points (2, 3) and (6, 9) can be shown to produce slope equal to $\frac{3}{2}$ or 1.50.

Constant functions can be introduced with the graphing calculator by comparing driving at different speeds. Consider, for example, the equations $Y_1 = 55x$, $Y_2 = 60x$, $Y_3 = 65x$, and $Y_4 = 450$. The graphical display of average speeds of 55 mph, 60 mph and 65 mph and the constant function (distance is 450 miles) where x is time and y is distance give students an everyday example of linear relations that can be investigated graphically. Using TRACE students can estimate different times and distances or compare the time that it takes to travel 450 miles.

We can use these real-world situations to investigate slope by graphing the three or four equations sequentially, allowing students to build intuition for linear relations. Once students understand the effect of the coefficient of x on the graph, we further explore the graphs of y = 2x, y = 2x + 1, and y = 2x - 2 and other graphs to round out our investigation of y = mx + b. Students can deal with abstraction more easily when they are led from specific real-world examples to generalizations.

Interpreting Graphs and Problem Solving

Pre-algebra students' ability to interpret linear and non-linear graphs and make predictions can be enhanced by using the graphing calculator to explore real-world examples using the TRACE function and the linear regression model. As an example (Kennedy, in press), suppose from a data set a scientist determines that

$$y = -\frac{1}{4}x^2 + \frac{15}{4}x + 25$$

closely approximates a monkey's performance, y, when various levels of anxiety, x, are measured. Students can explore the graph to determine at what level anxiety does the monkey perform best and when can the monkey no longer perform. They can then make other predictions about the monkey's performance.

Students can extend the work done with linear functions by examining real-world data. Data that seems to be linear can be collected to use the linear regression capability of the calculator that can allow students to make predictions about unobserved quantities. Consider the problem in Figure 6.

Biologists at Texas A&M have recorded the following data and claim that crickets chirp faster on warm days:

Chirps per minute	66	58	94	120	83	119	121	65	55
Temperature	55	53	62	70	59	69	69	55	52

Figure 6. *Cricket chirp data*

We can prepare the calculator to draw a scatter plot and produce a line of best fit. It is important in this case to discuss restricted domain with students. That is, crickets only chirp at certain temperatures. Under STAT DATA on the TI-81 input the data points $x_1 = 55$ $y_1 = 66$, $x_2 = 53$ $y_2 = 58$ and so on. Set RANGE for x from 30 to 80 by five and for y from 45 to 160 by twenty. Under STAT DRAW choose SCATTER and press ENTER (scatter plot appears). We can then choose Lin Reg and paste in the equation for the regression line. TRACEing now allows the students to make predictions about the cricket's behavior at different temperatures. In a pre-algebra statistics unit students can gather their own data using surveys and make similar comparisons and predictions.

From the examples above it is easy to see that the graphing calculator is an excellent tool for investigating many types of problems. If we extend the "box-problem" by generalizing the formula $y = (10-2x)(8-x)x$, students can investigate further the size square that yields the most volume by using TRACE and ZOOM alternately. It is clear that appropriate use of the graphing calculator enables students to explore higher mathematical ideas early on and develop intuition for the algebraic concepts that will follow.

Some Preliminary Research Results

At this reporting, PATH Mathematics is at the end of the pilot semester. The results reported here reflect the work done with three pilot classes of 60 ninth graders, though data are not available for all students. Students were recruited in January from existing Pre-Algebra and Consumer Mathematics classes. Consumer Mathematics is generally taken before or instead of Pre-Algebra.

A survey was given to all of the classes in the last week of the semester. Table 1 shows the results of the items that relate to calculator use. Some of the items are variations of an earlier survey (cited by Hembree & Dessart, 1992).

Table 1. *Survey Results (N=53)*

Item	Mean	% Strongly Agree/Agree
26. I used a graphing calculator on the last test (about lines) to solve problems and check my work.	2.28	70
27. I understand the graphing problems we have done on the graphing calculator.	1.92	88
28. I understand how to use the fraction calculator to solve fraction problems.	1.76	92
34. Using the graphing calculator helps me to learn algebra.	2.24	66
35. Calculators make math fun.	2.18	72
36. Math is easier if calculators are used to solve problems.	1.86	80
37. Since I use a calculator I do not need to know how to do math with paper and pencil.	3.94	14
38. Using a calculator sometimes seems like cheating.	2.76	56
40. I understand what slope and y-intercept mean.	1.96	86

Strongly Agree=1; Agree=2, Neutral=3, Disagree=4; Strongly Disagree=5

The results show that students have a positive attitude toward calculator use. Item 37 indicates that students still believe that it is important to be able to compute with paper and pencil. Students were evenly divided on Item 38: "Using a calculator sometimes seems like cheating." The word "sometimes" is important in the item. Students were allowed to use calculators on all tests, except in the case when they were asked to estimate or to model an operation with manipulatives. We expected that they would be quite comfortable with using calculators in general and not feel that they were cheating. This is a problematic attitude reported in other research. It will be interesting to investigate it further as PATH Mathematics continues. In the second year of the project the survey will be refined and given to treatment and control classes as a pre and post assessment.

In the last unit of the pilot semester, slope and the equation of the line were developed using simple real world problems and exploration on the graphing calculator as described in the previous section. This explains the reason for including Item 40 in the calculator block. A test was given to assess student understanding of slope, intercept, the equation of the line, and graphical solution to systems of equations. Students were encouraged to use graphing calculators throughout the test and were required to find a point of intersection with the calculator for one test item. Survey Item 26 refers to that calculator use. A statistically significant correlation ($p<.05$) between positive attitude and confidence in mathematics was noted. A highly significant correlation ($p=.003$) was found when comparing scores on the linear function test and the calculator block. Positive attitude and use of calculator appear to have a positive impact on performance in algebra.

One important finding of the current study is that there appears to be no correlation between arithmetic skill and student learning of algebra. In the first week of the semester, the Arithmetic Skills test from the Descriptive Tests of Mathematics Skills of the College Board series was given in all classes. In comparing student scores on the arithmetic test with performance on the algebra (linear function) test, a non-significant correlation ($r=-.06$) was noted. This indicates that, indeed, students with limited arithmetic skill can pursue higher mathematics, just as we suspected.

The research will continue into next year and beyond. PATH Mathematics classes will be compared with control classes next year through pre and post survey and content instruments. One interesting indicator of success will be student entry into and successful exit from higher level classes. The PATH students will be tracked through the next three years of their high school experience.

Conclusion

The graphing calculator is an excellent tool for problem solving and for developing intuition for algebraic ideas. Use of the calculator complements an inductive constructivist approach to the learning of algebra. Students who have been historically tracked out of algebra can gain access to higher mathematics through an enriched, high expectations, technology-intensive curriculum that integrates the graphing calculator throughout. Calculator technology can be used to remove the barrier of computation to allow students to explore ideas and solve problems.

References

Army, P. D. (1991). An approach to teaching a college course in trigonometry using applications and a graphing calculator. *Dissertation Abstracts International, 52,* 8A. (University Microfilms No. 92- 03038, 2850)

Becker, B. A. (1991). The concept of function: Misconceptions and remediation at the collegiate level (function concepts, precalculus). *Dissertation Abstracts International, 52,* 8A. (University Microfilms No. 92- 03039, 2850)

Charles, L. H. (1990). *Algebra thinking: First experiences.* Sunnyvale, CA: Creative Publications.

Educational Testing Service. (1979). *Descriptive tests of mathematics skills of the College Board: Arithmetic skills form B.* Princeton, NJ: Author.

Embse, C. (1992). Concept development and problem solving using graphing calculators in the middle school. In J. T. Fey (Ed.), *Calculators in mathematics education: 1992 yearbook* (pp. 65-78). Reston, VA: National Council of Teachers of Mathematics.

Fitzgerald, W., Winter, M. J., Lappan, G., & Phillips, E. (1986). *Middle grades mathematics project: Factors and multiples.* Menlo Park, CA: Addison-Wesley.

Giamati, C. M. (1990). The effect of graphing calculator use on students' understanding of variations on a family of equations and the transformations of their graphs (calculators). *Dissertation Abstracts International, 52,* 1A. (University Microfilms No. 91-16100)

Hembree, R., & Dessart, D. J. (1992). Research on calculators in mathematics education. In J. T. Fey (Ed.), *Calculators in mathematics education: 1992 yearbook* (pp. 23-31). Reston, VA: National Council of Teachers of Mathematics.

Kennedy, P. A. (1991). *Fundamentals of mathematics, pre-algebra, consumer mathematics: Mathematical staff development module 26.* Austin, TX: Texas Education Agency.

Kennedy, P. A. (in press). Investigating polynomials and systems of equations with the graphing calculator. *School Science and Mathematics.*

National Research Council. (1989). *Everybody counts: A report to the nation on the future of mathematics education.* Washington, DC: Mathematical Science Education Board.

Nichols, E. D., Litwiller, B. H., & Kennedy, P. A. (1992). *Holt pre-algebra.* Orlando, FL: Holt, Rinehart and Winston.

Post, T. R., Behr, M. J., & Lesh, R. (1988). Proportionality and the development of prealgebra understandings. In A. F. Coxford (Ed.), *The ideas of algebra K-12: 1988 yearbook* (pp. 78-90). Reston, VA: National Council of Teachers of Mathematics.

Rich, B. S. (1990). The effect of the use of graphing calculators on the learning of function concepts in precalculus mathematics. *Dissertation Abstracts International, 52,* 3A. (University Microfilms No. 91-12475, 835)

Vazquez, J. L. (1990). The effect on the calculator on student achievement in graphing linear functions. *Dissertation Abstracts International, 51,* 11A. (University Microfilms No. 91-06488, 3660)

The Calculator and Computer Precalculus Project (C^2PC): What Have We Learned in Ten Years?

Bert K. Waits
Franklin Demana
The Ohio State University

We have a philosophy about incorporating technology into instruction that is grounded in the experience gained from our ten year old Calculator and Computer Precalculus (C^2PC) project that we co-direct at Ohio State University. The project began as an extension of our experience and that of Joan Leitzel, Alan Osborne, and Joe Crosswhite in the Transitions to College Mathematics high school project that started at Ohio State in 1980 (Demana & Leitzel, 1988). The Transitions project was extended to include middle school in 1984 and called the Approaching Algebra Numerically (AAN) middle school project (Comstock & Demana, 1987). The Transitions, AAN, and C^2PC projects had their origins in an effort to reform the college remedial mathematics curriculum that began in 1974 at Ohio State and involved the use of four-function calculators by all students (Waits & Leitzel, 1976). Ohio State University was the first major U.S. college or university to give up teaching 18 year olds paper and pencil *arithmetic*. These were college "remedial" students who had failed to learn high school algebra and geometry. Calculators were required of all these students. They learned elementary and intermediate algebra in a new technology enhanced, supportive environment. Students who had failed with the traditional approach found success in this new setting: a setting supported by calculator generated numerical and graphical investigations. The AAN middle school project also used calculators to build a numerical and graphical foundation for the study of algebra. Interestingly, the hand-held calculator used in remedial mathematics courses at Ohio State in 1974 was the TI Exactra 19, a four function, 6 digit LED calculator. By the late seventies Ohio State was using the TI-30, a full featured scientific calculator, in all freshman mathematics courses including calculus.

Therefore, we had a natural interest in addressing a related non-remedial problem experienced at Ohio State regarding the mathematics preparation of calculus bound students. Many students entering Ohio State directly from high school with 3 or 4 years of college preparatory mathematics were not prepared for the study of calculus, or even "college algebra" (Demana & Waits, 1988a). We decided to develop a new technology enhanced high school precalculus course with the goal of increasing the number of students ready for university level calculus. The C^2PC development team (ourselves, Alan Osborne, Greg Foley, and five high school teachers) decided computer generated visualization would be our vehicle to enhance the teaching and learning of precalculus mathematics. The C^2PC project materials evolved over a six-year period at many pilot and field test sites both in high schools and in colleges. The project materials evolved into a textbook

(Demana, Waits, & Clemens, 1992) which is recognized as the first widely adopted high school or college textbook *to require* the fully integrated use of computer graphing technology *by students* on a regular basis. Several research studies have been conducted as an outgrowth of the C²PC project. Dunham (1992) nicely summarizes some of them, and Harvey and Osborne (in press) have analyzed the C²PC field test data from 1988.

We have trained many of our C²PC teachers in intensive one-week summer institutes. These teachers are our most valuable resource. Without them our project would have failed. We have grown tremendously in the past ten years because of the many exceptional classroom teachers with whom we have had the privilege to be associated. We have learned from them and continue to learn from them. We have learned that change occurs in small, incremental steps. This fact can't be over emphasized. We have learned the importance of taking a familiar body of high school material (in our case, fourth-year high school analysis, functions, theory of equations, analytic geometry, and trigonometry or "college algebra and trigonometry") and then making the assumption that computer graphing (function and parametric) would be available to *all* students on a regular basis for both in-class activities and homework.

We began out project with desktop computers in 1982 and developed our own graphing software (Demana & Waits, 1987-90). However, by 1989 we were almost a 100% graphing-calculator-driven project. We began using Casio *fx*-7000G graphing calculators in 1986 and now use the TI-81 almost exclusively. Frankly, the invention of graphing calculators by Casio is what made C²PC a viable and implementable course.

Students at typical high schools rarely have regular access to a computer lab during a precalculus class and a computer at home. In fact, most precalculus teachers we know indicate they seldom used computers and when they did, it was for "demonstration" only. They reported it was almost impossible to schedule their mathematics classes in a computer lab because the labs were fully scheduled with non-mathematics classes. Software presented additional problems. Some software required training that many teachers found difficult or inconvenient to obtain.

We believe that *students* must use computers on a regular basis for both in-class work and for homework outside of class if there are to be significant changes in the mathematics that students learn in the 1990's. In fact, the first Casio graphing calculator (Casio fx-7000G) was a computer with built-in graphing software! Modern graphing calculators like the TI-81 have many computer-like features. They have standard computer processors, display screens, built-in software, and they are fully programmable. They are *personal* computers that fit in a pocket or purse. The built-in graphing software on modern graphing calculators includes function graphing utilities and parametric graphing utilities that do not require programming and have simply amazing features. In addition, modern graphing calculators have statistical functionality with a graphics interface and powerful matrix arithmetic capabilities. Some now have I/O capability for sharing programs and files, like the TI-85 (Demana & Waits, 1992a).

The grade 9-12 standards in *Curriculum and Evaluation Standards for School Mathematics* (National Council of Teachers of Mathematics, 1989) includes the assumption that graphing calculators will be available to *all* students at *all* times. Our ten year C²PC experience with computer visualization points to the wisdom of this important assumption. Inexpensive graphing calculators make computer visualization for *all* students a realistic, obtainable goal.

The C²PC Experience

What did we do in C²PC that is different? Almost everything - almost nothing! The content is easily recognizable. However, the tools used to teach and learn the familiar content are new and different. We like to think of our C²PC course as an exciting, problem solving course that includes a study of traditional precalculus topics and also foreshadows the study of calculus.

We approached the incorporation of hand-held visualization technology as a natural evolution of our positive experience with scientific calculators. As the project evolved, we found we used computer (graphing calculator) visualization tools to do ten types of fundamental activities that occurred repeatedly in every chapter of our project materials and every day in our project classrooms. (Note: Every activity assumes the use of graphing calculators.)
1. Approach problems numerically.
2. Visually support the results of applying algebraic paper and pencil manipulations to solve equations and inequalities.
3. Use visual methods to solve equations and inequalities and then confirm using analytic algebraic paper and pencil methods.
4. Model, simulate and solve problem situations and confirm, when possible using analytic algebraic paper and pencil methods.
5. Use computer generated scenarios to illustrate mathematical concepts.
6. Use visual methods to solve equations and inequalities that *cannot* be solved using analytic algebraic methods.
7. Conduct mathematical experiments; make and test conjectures.
8. Study and classify the behavior of different classes of functions.
9. Foreshadow concepts of calculus.
10. Investigate and explore the various connections among different representations of a problem situation.

Illustrations

In this section we illustrate each of the ten major activities. All graphing calculator references are based on the TI-81 graphing calculator.

Activity 1. Approach problems numerically: the graphing calculator is the best *scientific* calculator students will ever use!

Example 1A: When will your money triple in value at 6% interest compounded annually?
Solution: Make a guess, say 15 years. Notice that the problem *and* solution are both visible in Figure 1.1 (left). You see only the answer on ordinary scientific

calculators. Seeing both the problem and solution at the same time is a powerful pedagogical feature of graphing calculators.

Figure 1.1. *Solving the equation $1.06^x=3$ by repeated guess and check on the computer screen of a graphing calculator*

Next we engage C²PC students in "math talk." For example, should the next guess be smaller or larger than 15 years? Figure 1.1 (middle) shows the result of making a second guess of 20. This process is enhanced by the TI-81 because pressing the up arrow key recalls the last *problem*. Then it is easy to edit and obtain another solution. Now two problems and two solutions are visible and students easily understand the solution is between 15 and 20 years. The guess and check process can be quickly repeated to obtain a very accurate solution as shown in the last screen of Figure 1.1.

Example 1B: Solve $x - \cos x = 0$ by recursion (fixed point iteration).
Solution: Figure 1.2 displays the recursive solution. It is obtained by simply using the STO key and repeated presses of the ENTER key.

```
1 → x
                    1
cos x → x
.5403023059
.8575532158
.6542897905
.7934803587
```

```
.7390851335
.739085133
.7390851334
.7390851331
.7390851333
.7390851332
.7390851332
```

Figure 1.2. *Finding a solution to the equation $\cos x - x = 0$ using recursion*

In these two examples we used the computer screen to do the calculations with both the problem and answer displayed. This powerful pedagogical feature also gave us the opportunity to confront the issues of estimation (a new basic skill), efficiency of a "guess and check" method, error, and exact versus approximate answers. For example, consider the *barrier* the "exact" solution for Example 1A (namely $\frac{\log 3}{\log 1.06}$) presents to students. Just what is $\frac{\log 3}{\log 1.06}$? Of course, we approximate this number with a calculator when used in a real problem. Also we challenge you to try to find the "exact" solution to $\cos x - x = 0$.

Exactness is over emphasized in school mathematics. Mathematics teachers (secondary and collegiate) are virtually the only ones who find beauty in exact numbers such as $\frac{1+\sqrt{2}}{3}$. Almost everyone who uses and values mathematics will immediately approximate this number. Students need to understand that it is *correct* to use approximations. The word "solve" in school mathematics needs to be redefined to mean "determine the solution with a prescribed degree of accuracy" rather than "find the exact solution" (Demana & Waits, 1992b).

Activity 2. Visually support paper and pencil algebra processing.

Example 2: Solve the inequality $x^3 - 4x < 0$.
Solution: By factoring and applying a sign pattern method, the solution is $(-\infty, -2) \cup (0, 2)$. Our students make the connection that the solution of $x^3 - 4x < 0$ is also given by the values of x for which the graph of $y = x^3 - 4x$ is *below* the x-axis. Thus the graph in Figure 2.1 provides support for the analytically determined answer. The notation [-4, 4] by [-10, 10] below the figure describes the rectangular "viewing window" determined by $-4 \leq x \leq 4$ and $-10 \leq y \leq 10$ which is set on most graphing calculators by using the RANGE key.

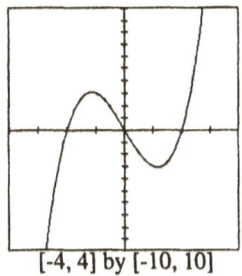

[-4, 4] by [-10, 10]

Figure 2.1. *A complete graph of* $y = x^3 - 4x$, *all below the x-axis for x in the interval* $(-\infty, -2)$ *or* $(0, 2)$

This type of activity empowered C²PC students with the ability to make connections among the algebraic and geometric representations of the "problem" (in this case the inequality) and support (check) their solutions found analytically, visually. We found that the bread and butter of C²PC activities centered around "do algebraically and support graphically."

Activity 3. Use a graphing calculator to visually <u>do</u> the algebra processing (thus replacing the "need" for some algebraic manipulation). Confirm using analytic methods.

Example 3: (Use a graph to) solve $6x^2 + 7x - 25 = 0$. Confirm using the quadratic formula.
Solution: A complete graph in Figure 3.1(a) shows there are two real solutions at about -2.8 and 1.5. Zoom-in can be used to quickly obtain highly accurate solutions. The positive solution is shown in a magnified view in Figure 3.1(b). The

quadratic formula yields exact solutions of $\dfrac{-7+\sqrt{649}}{12}$ and $\dfrac{-7-\sqrt{649}}{12}$. The positive solution is 1.539623200 (to 10 digits). The TRACE value in Figure 3.1 (b) is accurate to five digits.

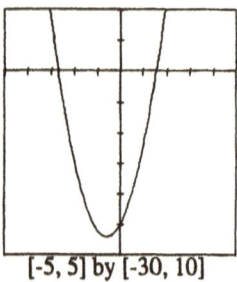
[-5, 5] by [-30, 10]

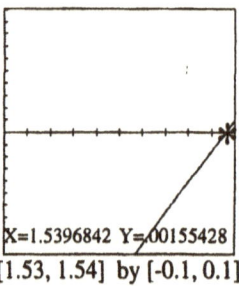
X=1.5396842 Y=.00155428
[1.53, 1.54] by [-0.1, 0.1]

Figure 3.1. *(a) Complete graph of* $y = 6x^2 + 7x - 25$ *and (b) a zoom-in view with a TRACE value of the positive x-intercept (the positive solution of* $6x^2 + 7x - 25 = 0$ *)*

This activity gave our students new "tools" (visually applied and computer generated) to do traditional algebraic processing activities. We then added the important activity of *analytic confirmation* using manipulative algebra. This is *very* important because we are in a transition decade. The mathematics and scientific community have not yet agreed on a minimal list of paper and pencil algebraic skills. The visual activity followed by paper and pencil analytic methods (algebra manipulation) is a wonderful intermediate step. We believe this approach actually strengthens students ability to do paper and pencil manipulative algebra (if that's what's valued). In the future, the "community" will come to accept computer graphing or computer algebra technology as a *tool* to "solve, factor, etc." just as the scientific calculator is now accepted as a *tool* to compute and evaluate numerical expressions. We personally can confirm that the precalculus curriculum we teach has changed dramatically in the past 30 years. And we are certainly due for even more dramatic changes in this decade because of graphing calculators!

Activity 4. Model and/or simulate a problem situation (application) and solve using visualization tools. Confirm, if possible or appropriate, using analytic methods.

Example 4: A tanker on lake Erie leaves Toledo, Ohio, at 8:00 AM and heads directly for Buffalo, New York. A passenger ferry leaves Buffalo, at 10:00 AM and heads directly for Toledo. Assume the tanker is making a steady 17 mph, the ferry is making 23 mph, and Toledo and Buffalo are 176 miles apart. When will the two boats pass and at what distance from Buffalo? Simulate and solve this problem situation then confirm algebraically (using paper and pencil).
Solution: This is an activity using the parametric graphing utility built-in on modern graphing calculators. Set the following:
   X1T = 17T   Y1T = 1
   X2T = 176 - 23(T - 2)   Y2T = 1.1
X1T and X2T are the algebraic representations of the position of the tanker and ferry in terms of time T. Run the simulation setting Tmin = 0, to Tmax = 12

(hours), Tstep =0.1 (1/10th. of an hour), with SIMUL(taneous) mode selected. Use the viewing window shown in Figure 4.1. The motion and position of the tanker and ferry become visible and real! The static screens shown in Figure 4.1 do not do justice to this dynamic simulation.

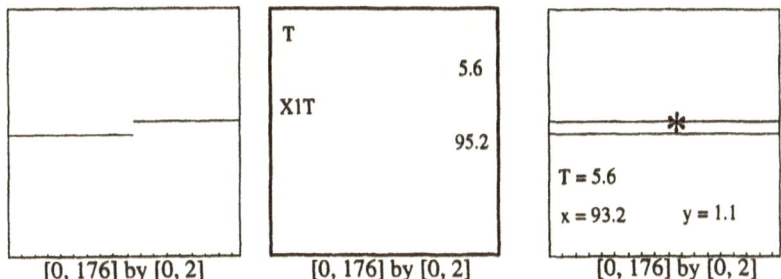

Figure 4.1. *A simulation of the Lake Erie tanker from Toledo passing the Lake Erie ferry from Buffalo*

The TI-81 simulation can be interrupted when they "meet" by pressing the ON key (as shown in the left screen in the figure) and then the value of time T and position X1T can be obtained from memory (middle screen). Finally the TRACE key can also be used to "see" the position of the ferry and tanker at various times T.

It is easy to see that the tanker and ferry pass at about 1:30 p.m. $\left(8 + 5\frac{1}{2}\right)$ when they are about 82 miles from Buffalo. Analytic methods can be used to show they pass exactly when $T = 5.55$ and they are 81.65 miles from Buffalo at that time.

C^2PC students are challenged to find the (exact) solution using paper and pencil analytic methods (very easy in this case). Our experience indicates that they go through the algebraic manipulations with far greater interest and enthusiasm when the analytic work is embedded in interesting activities like this simulation! This type of computer aided visual simulation also adds a very exciting dimension to the teaching and learning process. We saw our project teachers get excited and they, in turn, excited their students. Parametric graphing is almost unknown in today's school mathematics curriculum, but it will have an increasingly important role in the future (Demana & Waits, 1989).

Activity 5. Illustrate mathematical ideas and applications (in pedagogically powerful geometric settings).

Example 5: Show the relationship between the circular sine function definition and the graph of $y = \sin x$ (unwrapping the wrapping function).
Solution: This is a demonstration without words! Select parametric and simultaneous graphing mode. Set the following:
 X1T = -1 + cos T   Y1T = sin T
 X2T = T   Y2T = sin T

Run this demonstration from Tmin = 0 to Tmax = 6.28 with Tstep = 0.1 in the TI-81 square viewing window [-2, 6.28] by [-2.76, 2.76]. TRACE to see values on both the circle and the sine curve.

You must try Example 5 if you have not experienced this type of powerful, dynamic visualization. Try repeating this activity with Y2T = cos T or tan T. Again the static screens shown in Figure 5.1 do not convey the real pedagogic value of simulation. (Notice that the square window used to construct Figure 5.1 is not the same as the TI square window, because we are using a different grapher to construct the figures for this paper.)

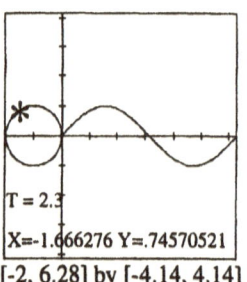
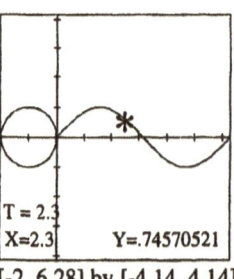

[-2, 6.28] by [-4.14, 4.14]   [-2, 6.28] by [-4.14, 4.14]

Figure 5.1. *Unwrapping the circular sine function using parametric graphing*

These types of activities provide dynamic, visual insight and understanding for students in ways not possible with paper and pencil methods alone. And each individual student can conduct this "demonstration."

Activity 6. Solve easily understandable and "real" problems that algebra students *cannot* solve with paper and pencil algebraic manipulation.

Example 6: A box *with lid* is constructed from an 8.5 inch by 11 inch piece of cardboard by removing four squares and two *rectangles* as shown in Figure 6.1.
 1. Write an algebraic representation of the volume of the box in terms of x (x given in Figure 6.1 is the side length of the removed squares)
 2. Draw a complete graph of the algebraic representation.
 3. What values of x make sense in this problem situation?
 4. Draw a graph of the problem situation.
 5. Determine the sizes of squares and rectangles to remove that produce a box with lid of volume equal to 25 cubic inches.
 6. Determine the sizes of squares and rectangles to remove that produce a box of maximum volume. What is the maximum volume?

Solution: The algebraic representation is $V = x(8.5 - 2x)(5.5 - x)$. Here the key is for the student to recognize that V = L*W*H and H = x, W = 8.5 - 2x and L = y. Then the relationship $2x + 2y = 11$ is used to determine y. All of this activity is deduced from a real model (cut out of paper) and Figure 6.1. The graph in Figure 6.2(a) is a complete graph of the algebraic representation. Notice there are *three* solutions to V = 25 but one of them is extraneous to the "real" problem situation. Figure 6.2(b) shows that when x is about 0.75 or 2.55, the resulting box with lid has volume about 25 more accurate results can be obtained by zooming. The last screen shows that the volume is maximum (about 33.074 cubic inches) when x is about 1.585 (inches). These last numbers are stated with error less than 0.01. Do you see how the error can be deduced from Figure 6.2(c), looking carefully at the dimensions of the window? This kind of error analysis is an important activity in C²PC. It should be noted that no student (or teacher for that matter) could find the "exact" solution to V=25 with paper and pencil algebra. And only calculus students could find the "exact" solution to the maximum volume problem.

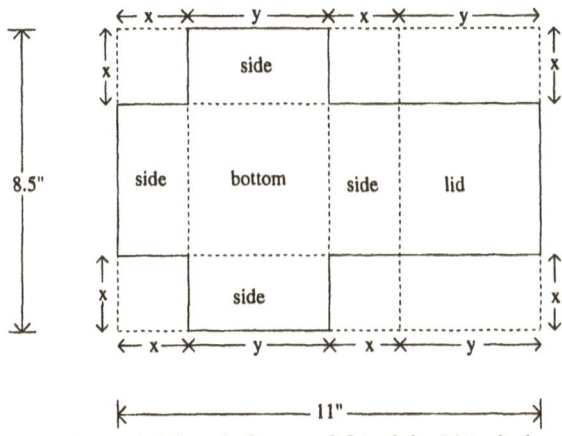

Figure 6.1. A box <u>with lid</u> made from an 8.5 inch by 11 inch sheet of cardboard by removing four x inch squares and two x inch by $(x + y)$ inch rectangles where $2x + 2y = 11$ or $y = 5.5 - x$

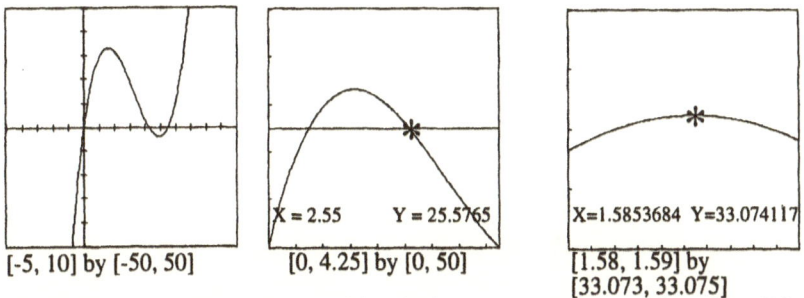

Figure 6.2. (a) A complete graph of the algebraic representation of the volume V in terms of x, (b) One TRACE solution (estimate) of a value of x (2.55) that produces a box with lid of volume about 25, and (c) A highly magnified view of the local maximum value of V

Activity 7. Conduct mathematical experiments; make and test mathematical conjectures

Example 7A: Graph $y = \sin x + \cos x$. Is it a sinusoid? Is so, what is its equation? Is $y = 2\sin x + 3\cos x$ a sinusoid? Conjecture a generalization. Prove your generalization.

Solution: A sinusoid is any graph of the form $y = A + B \sin(Cx + D)$: a horizontal or vertical stretch or shrink (dilation) and/or horizontal or vertical translation (slide) of $y = \sin x$. A complete graph shows that $y = \sin x + \cos x$ is certainly a sinusoid. What are the values of A, B, C, and D? Here C^2PC students *explore and investigate* as illustrated in Figure 7.1. From the graphic analysis it is easy to conjecture that A = 0, B = $\sqrt{2}$, C = 1 and D = $\pi/4$. This is a conjecture that can be supported by graphing $y = \sqrt{2} \sin(x + \pi/4)$ and $y = \sin x + \cos x$ in the same window.

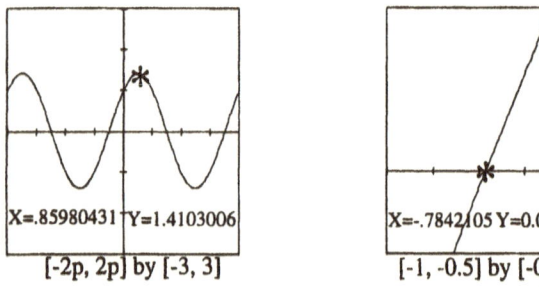

Figure 7.1. *Graphs of $y = \sin x + \cos x$ showing the high point $\left(\sqrt{2}\right)$ and a magnified view of the largest negative x-intercept ($-\pi/4$)*

The graphs appear to be the same. This is not a proof. C^2PC students now use analytic methods to *prove* that $\sqrt{2} \sin(x + \pi/4) = \sin x + \cos x$ by expanding $\sqrt{2} \sin(x + \pi/4)$ using the $\sin(\alpha + \beta)$ sum formula from trigonometry. Again, notice the wonderful marriage between graphic and analytic methods. We then make the problem a bit more difficult by asking the sinusoid question for $y = 2\sin x + 3\cos x$ (Figure 7.2).

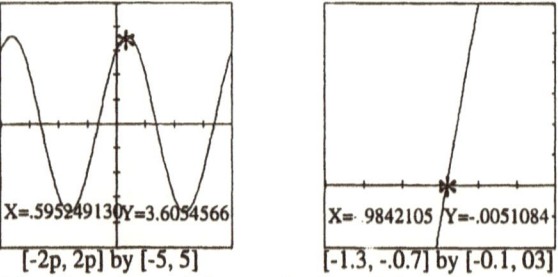

Figure 7.2. *Two views of the graph of $y = 2\sin x + 3\cos x$. Is it the sinusoid $y = \sqrt{13} \sin(x + \alpha)$ where $\alpha = \tan^{-1}\frac{3}{2}$?*

Our students are led to conjecture that $y = A\sin x + B\cos x$ is the sinusoid $y = \sqrt{A^2 + B^2} \sin(x + \alpha)$ where $\alpha = \tan^{-1}\frac{B}{A}$. The proof is again a direct application of the sine sum of two angles formula.

Finally we conclude this type of exploration with the sinusoid question for $y = 2\sin x + \cos 3x$. It is *not* a sinusoid as a complete graph will easily show! Electrical engineers have a neat way of looking at $2\sin x + 3\cos x$ summarized in Figure 7.3.

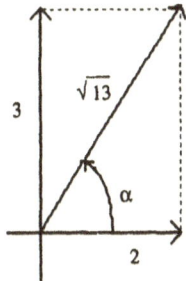

Figure 7.3 . *An electrical engineer's way of thinking about* $y = 2\sin x + 3\cos x$, *thinking of* $\sin x$ *as the unit vector* $i = (1,0)$ *and* $\cos x$ *as the unit vector with the sum vector* $v = 2(\sin x) + 3(\cos x)$ *written as the form of its magnitude,* $|v|$, *and direction,* $\frac{v}{|v|}$

Example 7B: Devise a method of factoring *any* cubic polynomial with real coefficients.
Solution: As an example of the procedure we consider an arbitrary cubic polynomial, say $5x^3 - 10x^2 + 17x - 23$. The method presented here works for any cubic polynomial. First determine a complete graph of $y = 5x^3 - 10x^2 + 17x - 23$. It has at least one real root (Why?). Zoom-in on one of the real roots, say $x = R$ (Figure 7.4).

R is computed to as much accuracy as desired (up to about 9 or 10 digits on the TI-81). It follows that $x - R$ is a factor. Now we graph the function $g(x) = \frac{y}{x - R}$. Notice it is quadratic in behavior (Figure 7.5). Of course this is as it should be! Next we zoom-in on the vertex of the parabola. Call it (A, B). Again A and B may be found to a high degree of accuracy by zooming. Finally we write $5x^3 - 10x^2 + 17x - 23 = (x - R)[5(x - A)^2 + B]$ and complete the factorization by applying the quadratic formula. C²PC students *know* that R, A and B exist and, perhaps the more important practical issue is that, they know *how* to find them! Consider the wonderful connections our C²PC students establish among the old theory of equations topics and modern graphing technology through this process of factorization (Figure 7.6). The fundamental theorem of algebra comes alive for C²PC students!

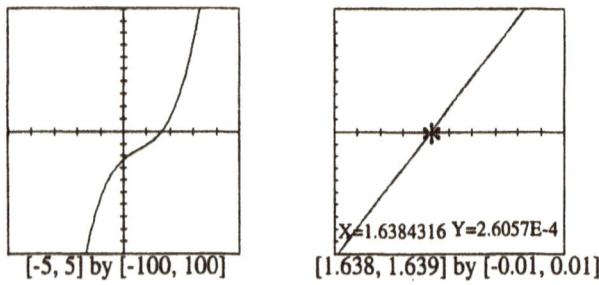

Figure 7.4. *A complete graph of* $y = 5x^3 - 10x^2 + 17x - 23$ *and a highly magnified view of its x-intercept*

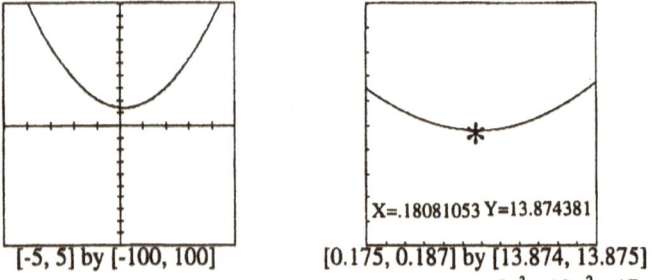

Figure 7.5. *A complete graph of the parabola* $g(x) = \dfrac{5x^3 - 10x^2 + 17x - 23}{x - R}$ *and a magnified view of its vertex*

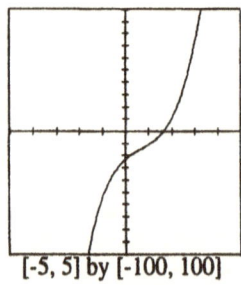

Figure 7.6. *Apparently identical graphs of* $y = 5x^3 - 10x^2 + 17x - 23$ *and* $y = (x - R)\left[5(x - A)^2 + B\right]$ *with R as the x-intercept of* $y = 5x^3 - 10x^2 + 17x - 23$ *and (A, B) is the vertex of the parabola* $g(x) = \dfrac{y}{x - R}$ *from Figure 7.5*

CALCULATOR AND COMPUTER PRECALCULUS PROJECT

Activity 8. Investigate properties of common classes of functions (polynomials, rationals, radicals, exponentials, trigonometric, etc.).

Example 8: Discuss the asymptotes of rational functions of the form $y = \dfrac{N(x)}{D(x)}$ where N and D are polynomials of degree 1, 2 or 3.

Solution: Degree 1 in both numerator and denominator leads to the general form: $y = \dfrac{Ax + B}{Cx + D}$, $C \neq 0$. By graphing many examples (Figure 8.1, left) the students make the connections between the analytic representation (the vertical asymptote $x = -D/C$) and its visual interpretation. Similarly, the analytic horizontal asymptote $y = A/C$ is associated with the flat shape of the graph for |x| large (Figure 8.1, right). Both asymptotes (lines) $y = \dfrac{A}{C}$ and $x = -\dfrac{D}{C}$ can be added to the graphs for visual support.

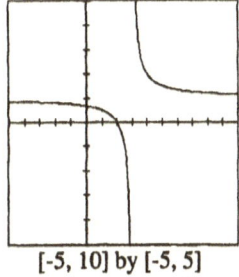
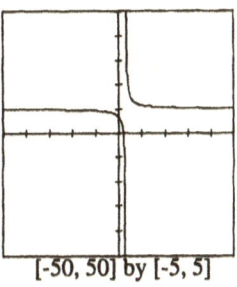

[-5, 10] by [-5, 5]   [-50, 50] by [-5, 5]

Figure 8.1. *Two views of* $y = \dfrac{x-2}{x-3}$ *showing vertical and horizontal asymptotes*

Figure 8.2 shows a typical case for a rational function with numerator degree 3 and denominator degree 1. Notice the graphical end behavior analysis leads naturally to a new concept of *end behavior model* ($y = x^2$ is an end behavior model for $y = \dfrac{x^3 - 10x^2 + x + 50}{x - 2}$). This is the higher level mathematics concept of *global analysis*. Furthermore, the polynomial division algorithm leads to the definition of a unique *end behavior asymptote*. From the division algorithm, $\dfrac{x^3 - 10x^2 + x + 50}{x - 2} = x^2 - 8x - 15 + \left(\dfrac{20}{x-2}\right)$. This purely analytic activity has a wonderful geometric interpretation shown in Figure 8.2(c). Notice how the quadratic graph most closely "matches" the rational function for all values of x except those near 2 (Demana & Waits, 1990a).

Here students begin to build their own intuitive understanding of the behavior of different classes of important functions. The rich intuition developed in precalculus aids in the analytic study of functions in calculus. This is constructivism at its best!

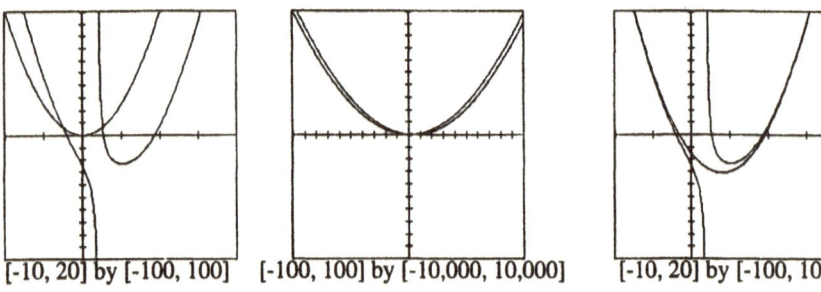

[-10, 20] by [-100, 100]   [-100, 100] by [-10,000, 10,000]   [-10, 20] by [-100, 100]

Figure 8.2. (a) Graph of $y = x^2$, (b) Graph of $y = \dfrac{x^3 - 10x^2 + x + 50}{x - 2}$, and (c) Graph of unique quadratic asymptote $y = x^2 - 8x - 15$ and $y = \dfrac{x^3 - 10x^2 + x + 50}{x - 2}$

Activity 9. Foreshadow the concepts of calculus.

Example 9: Investigate the limit of $y = \left(1 + \dfrac{1}{x}\right)^x$ as $x \to \infty$. Note: we ask this as a non-calculus question by asking C²PC students to find an end behavior model for the function.

Solution: The concept of limits arises naturally when using graphing calculators. Viewing the graph of $y = \left(1 + \dfrac{1}{x}\right)^x$ in a window longer than it is wide leads to the view in Figure 9.1 (right). It is easy to "see" the behavior of y as $|x|$ approaches ∞ (namely y approaches e, the horizontal line $y = 2.718$). Graphs make the limit concept intuitive and natural.

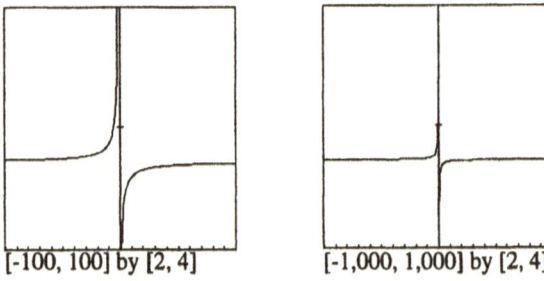

[-100, 100] by [2, 4]   [-1,000, 1,000] by [2, 4]

Figure 9.1. *Two views of the graph of $y = \left(1 + \dfrac{1}{x}\right)^x$ showing a horizontal asymptote (the line $y = e$)*

Other calculus concepts such as continuity and tangent lines arise naturally when studying functions with a graphing calculator. For example, we frequently encounter the important idea of "local straightness" and the idea that the tangent line is horizontal at a relative minimum or maximum. These ideas are illustrated every time graphing calculator zoom-in is used.

Activity 10. The development of multiple representations (including numeric, algebraic, matrix, and geometric) of a problem situation and the exploration of the connections among the different representations of the problem situation.

Example 10. Determine a fifth degree polynomial that passes through six different, but arbitrary points, say the six points (1, 8), (3, 4), (4, 6), (6, 8), (7, 1), and (9, 5) shown in Figure 10.1.

Solution: Let $y = Ax^5 + Bx^4 + Cx^3 + Dx^2 + Ex + F$ be a fifth degree polynomial that passes through the six given points. This leads to (by simple substitution) the following six *linear* equations with six unknowns A, B, C, D, E, and F. Try solving this system with paper and pencil! Of course, no one would even consider this today. This is a job for technology.

$$\begin{array}{rl} A+ \quad B+ \quad C+ \quad D+ \quad E+ \quad F & = 8 \\ 3^5A+ \ 3^4B+ \ 3^3C+ \ 3^2D+ \ 3E+ \ F & = 4 \\ 4^5A+ \ 4^4B+ \ 4^3C+ \ 4^2D+ \ 4E+ \ F & = 6 \\ 6^5A+ \ 6^4B+ \ 6^3C+ \ 6^2D+ \ 6E+ \ F & = 8 \\ 7^5A+ \ 7^4B+ \ 7^3C+ \ 7^2C+ \ 7C+ \ F & = 1 \\ 9^5A+ \ 9^4B+ \ 9^3C+ \ 9^2C+ \ 9C+ \ F & = 5 \end{array}$$

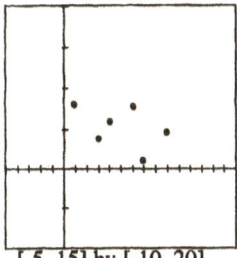

[-5, 15] by [-10, 20]

Figure 10.1. *A plot of the six points (1, 8), (3, 4), (4, 6), (6, 8), (7, 1), and (9, 5)*

Our C²PC students learn this 6 by 6 system is equivalent to a *matrix* representation given by:

$$\begin{bmatrix} 1 & 1 & 1 & 1 & 1 & 1 \\ 243 & 81 & 27 & 9 & 3 & 1 \\ 1{,}024 & 256 & 64 & 16 & 4 & 1 \\ 7{,}776 & 1{,}296 & 216 & 36 & 6 & 1 \\ 16{,}807 & 2{,}401 & 343 & 49 & 7 & 1 \\ 59{,}049 & 6{,}561 & 729 & 81 & 9 & 1 \end{bmatrix} \begin{bmatrix} A \\ B \\ C \\ D \\ E \\ F \end{bmatrix} = \begin{bmatrix} 8 \\ 4 \\ 6 \\ 8 \\ 1 \\ 5 \end{bmatrix}$$

or in more compact notation, $[A][X]=[B]$, where $[A]$ is the 6 by 6 matrix of coefficients, X is the column matrix of 6 unknowns A, B, C, D, E, and F and $[B]$ is the column matrix of constants 8, 4, 6, 7, 1, 5. It follows that the solution is given by the matrix product $[X] = [A]^{-1}[B]$ where $[A]^{-1}$ is the inverse of the coefficient matrix [A] (if it exists). Computing inverses and multiplying matrices are easy jobs for modern graphing calculators such as the TI-81. A quick calculator computation and multiplication yields:

$$[X] = \begin{bmatrix} A \\ B \\ C \\ D \\ E \\ F \end{bmatrix} = [A]^{-1}[B] = \begin{bmatrix} 0.0409722222 \\ -0.9020833331 \\ 7.010416664 \\ -23.17013888 \\ 30.670833332 \\ -5.649999992 \end{bmatrix}$$

Each entry of the solution matrix is accurate to 10 digits on the TI-81 and can be electronically stored to the proper coefficient in the 5th degree polynomial $y = Ax^5 + Bx^4 + Cx^3 + Dx^2 + Ex + F$! Try this TI-81 sequence of operations: $[A]^{-1}[B] \to [C]$, $[C](1,1) \to A$, $[C](2,1) \to B$, etc. Then the graph of $y = Ax^5 + Bx^4 + Cx^3 + Dx^2 + Ex + F$ can be added to the graph of the six points using the DRAWF command (Figure 10.2) for a very powerful and dramatic effect.

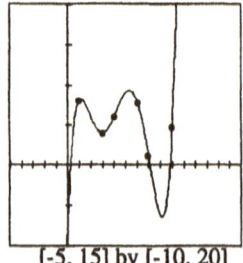

[-5, 15] by [-10, 20]

Figure 10.2. *A graph passing through the six points from Figure 10.1 with* $y = 0.04097x^5 - 0.90208x^4 + 7.01041x^3 - 23.17013x^2 + 30.67083x - 5.64999$

Think of the many connections that have been established with this example among the numeric (data), algebraic (6 by 6 linear system of equations), matrix (matrix representation of a 6 by 6 system), and geometric (the graph of the actual polynomial and graph of the data) representations. It is interesting to note that we began with a geometric representation (the visualized data points) and concluded with another geometric representation (the polynomial passing through the six data points). Other connections could be exploited as well (e.g. statistics).

What have we learned from C²PC?

We learned that new mathematical ideas are needed when graphing technology is used. Three of the new ideas were (a) viewing window, (b) complete graph, and (c) end behavior model. Before graphing technology was used we were content to graph (sketch) in an *unbounded* window as typically denoted in Figure 11.1.

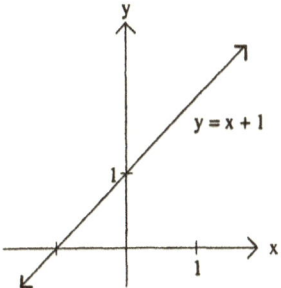

Figure 11.1. *A sketch of the line y = x+ 1*

However a graphing utility *requires* a bound on the values of x and y. Thus our students, and teachers had to learn to first specify a window xmin $\leq x \leq$ xmax, ymin $\leq y \leq$ ymax. This is typically done on graphing calculators on the RANGE edit screen. This viewing window feature led naturally to interesting problems, scale issues, and other ideas, as well as pitfalls discussed later.

The concept of a complete graph was necessary to answer the question, "How do you know you have found *all* the solutions when using a graphing method (or even Newton's method)?" For example, how do you know $y = x^3 - 5x^2 + 5x - 6$ has only one real solution when graphed in the [-10, 10] by [-10, 10] window (Figure 11.2)?

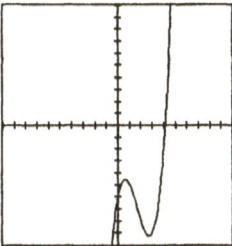

Figure 11.2. *A complete graph of* $y = x^3 - 5x^2 + 5x - 6$ *in [-10, 10] by [-10, 10]*

Why can't there be a solution for $x > 10$? The answer to this important question is that the displayed graph is "complete." That is, there is no other "important" behavior missing from Figure 11.2, in this case, no missing x-intercepts. The key is to build an intuitive understanding of what is considered "important" (intercepts, high and low points, increasing, decreasing, and end behavior) about different classes of functions. One modern role of calculus is to prove analytically when a graph is complete.

Perhaps the most important thing we learned in addition to the principle of incremental change was that algebra has a new role in school mathematics, as a *language of representation*, rather than as a tool for paper and pencil manipulation. It is clear to us that what is important today is the students' ability to correctly "algebraically represent" when confronted with a "real problem" rather than apply paper and pencil manipulations to a series of routine typical textbook problems. Other new ideas were "tool related" (e.g., screen coordinates vs. math coordinates).

We found that a technology enhanced graphing approach also made some old topics such as scale and error far more important (Demana & Waits, 1990b). We also had to confront many pitfalls (teaching opportunities!) such as hidden behavior (Demana & Waits, 1988b). One of the modern roles of differential calculus is to predict and locate behavior hidden from view. Just for fun, graph $y = \sin(60x)$ in [-10, 10] by [-3, 3]. Is the displayed graph correct?

We learned that the teaching and learning process changes when all students use hand-held technology. We observed a significant change in our teaching behavior and the behavior of our project teachers. We now lecture much less. We see the teacher as a guide or advisor. C^2PC students *actively* investigate, explore, make and test conjectures, and solve real problems. We learned the value of group or cooperative learning especially with high school seniors and college freshman. Our teachers reported that students' interest in mathematics and mathematics communication skills increased. Talking about mathematics became common!

How Do We Know C^2PC Works?

The fact is that we know the mathematics curriculum today must change because of technology. Mathematics is alive and changing and technology is part of that change. Appropriate use of technology promotes better mathematics teaching and learning. Tom Romberg says we are teaching 13th century arithmetic, 18th century analysis, and BC geometry in our schools today. He is correct! We MUST start teaching modern, applicable mathematics. We MUST use technology to help students *do* mathematics. Do C^2PC students do better on traditional paper and pencil algebra exams? We really don't care. Paper and pencil algebra skills are simply not very important today. What is important are the kinds of activities and learning that are illustrated in the ten examples given earlier where technology is a partner in the learning process.

New assessment methods are needed. New studies are needed on just what it is that C^2PC is doing. What are C^2PC students learning? Is it what we think? We don't have the answers. Much research is needed. Hopefully this research will be conducted world-wide as graphing technology use spreads. One very positive C^2PC evaluation is antidotal. Every AP calculus teacher that we have talked with (at least 30) claims that their C^2PC trained students do as well (and many better than expected) on the AP Calculus exams (with NO technology allowed) when compared with their past classes before C^2PC. It should be noted that many of these AP teachers are not C^2PC teachers and are somewhat unbiased. Also many report that when AP classes have mixed students (some from C^2PC, others from non-graphing calculator classes), they can readily identify C^2PC students as the ones having better problem solving and communication skills.

We do know the C^2PC approach "feels" better. We see students learning to value mathematics in ways we only dreamed of in the past. We see how a technology enhanced approach can be used to mathematically empower more students. Manipulative algebra is seen as less important in a new graphing technology intensive environment. Rich problem situations can be satisfactorily investigated by many more students. This is important in view of the *Standards* call for the same core mathematics content for *all* students.

We learned that the cost of graphing calculators is only a real issue with those who do not believe in their instructional use. C^2PC teachers learned to make parents and administrators partners in getting students the necessary graphing tools. Teachers held "open houses" for parents that included student demonstrations. These excited parents were far more willing to buy their children graphing calculators. C^2PC teachers came up with many other creative ways to obtain graphing calculators for their students.

Curriculum Reform Advice

The important thing to note from our C^2PC experience it that as we introduced the new activities (regular use of graphing calculators) we used the "old familiar curriculum." Teachers felt more comfortable first dealing with familiar topics and then dealing with wonderful new opportunities that exist in a graphing calculator environment. Some teachers told us that the first year through the C^2PC textbook they operated at the Examples 1 - 3 level almost exclusively. The second year is when they said it became "fun" (their words!) because they felt confident enough to move ahead and take "ownership" of the C^2PC course and move to incorporate more of the types of activities illustrated in Examples 4 - 10. Another point from our C^2PC experience is the recognition that teacher inservice is essential and can not be done in several afternoon workshops. Teachers are our most valuable resource, and we must provide for adequate training and follow-up.

Clearly five to ten years in the future, high school precalculus (and all of school mathematics) will change drastically to reflect the *Standards*. We believe pocket computer technology will be recognized as the major change agent which empowered thousands of dedicated school teachers. Graphing calculators can make the study of mathematics fun, can provide students with excellent learning experiences, and can provide a vehicle for *all* students to engage in doing real mathematics.

References

Comstock, M., & Demana, F. (1987). The calculator is a problem-solving concept developer. *Arithmetic Teacher, 34*(6), 48-51.

Demana, F., & Leitzel, J. R. (1988). Establishing fundamental concepts through numerical problem solving. In A. F. Coxford & A. P. Shulte (Eds.), *The ideas of algebra, K-12: 1988 yearbook* (pp. 61-88). Reston VA: National Council of Teachers of Mathematics.

Demana, F., & Waits, B. K (1987, 1988, 1989, 1990). *Master Grapher* (version 1.0) [computer software]. Reading, MA: Addison-Wesley.

Demana, F., & Waits, B. K. (1988a). Is three years enough? *Mathematics Teacher, 81*, 11-14.

Demana, F., & Waits, B. K. (1988b). Pitfalls in graphical computation, or why a single graph isn't enough. *College Mathematics Journal, 19,* 177-183.

Demana, F., & Waits, B. K. (1989). Around the sun in a graphing calculator. *Mathematics Teacher, 82,* 546-550.

Demana, F., & Waits, B. K. (1990a). Enhancing mathematics teaching and learning through technology. In T. J. Cooney & C. R. Hirsch (Eds.), *Teaching and learning mathematics in the 1990s: 1990 yearbook* (pp. 212-222). Reston, VA: National Council of Teachers of Mathematics.

Demana, F., & Waits, B. K. (1990b). The role of technology in teaching mathematics. *Mathematics Teacher, 82,* 27-31.

Demana, F., & Waits, B. K. (1992a). A computer for all students. *Mathematics Teacher, 84,* 94-95.

Demana, F., & Waits, B. K. (1992b). A case against computer symbolic manipulation in school mathematics today. *Mathematics Teacher, 84,* 180-183.

Demana, F., Waits, B. K., & Clemens, S. (1992). *Precalculus mathematics: A graphing approach* (2nd ed.). Reading, MA: Addison-Wesley.

Dunham, P. (1992). Teaching with graphing calculators: A survey of research on graphing technology. In *Proceedings of the fourth Conference on Technology in Collegiate Mathematics* (pp. 89-101). Reading, MA: Addison-Wesley.

Harvey, J. G., & Osborne, A. (in preparation). *Analysis of C^2PC field test data.* Paper in preparation.

National Council of Teachers of Mathematics. (1989). *Curriculum and evaluation standards for school mathematics.* Reston, VA: Author.

Waits, B. K., & Leitzel, J. (1976). Hand-held calculators in the freshman mathematics classroom. *American Mathematical Monthly, 83,* 731-733.

Supercalculators in Undergraduate Mathematics

Donald R. LaTorre
Clemson University

Clemson University has implemented a strong, broad-based initiative that effectively incorporates the HP-48S/SX supercalculator into its calculus, differential equations and linear algebra sequence. During each of the 1991-92 and 1992-93 academic years, over 50 class sections (approximately one-half of Clemson's offering of these courses) received significant calculator enhancement on a regular basis. Clemson has required the high level HP units in selected sections of calculus I, calculus II, multivariable calculus, differential equations and linear algebra since 1990-1991 and the students are responding enthusiastically.

Up until 1988, computing in Clemson's undergraduate mathematics courses had been restricted to writing or using programs for numerical calculations. But the new and emerging breed of high-level calculators, first evidenced by the Hewlett-Packard HP-28S, gave us the power to do much more: symbolic algebra, sophisticated graphics, variable operating modes, all in a mobile, active learning environment. Thus the project was conceived in response to a need to establish that such calculators (a) can be effectively integrated into the undergraduate mathematics curriculum and (b) can offer special advantages that are not readily available with mainframe or microcomputing technology.

The project has made a major impact on the Clemson campus: from small beginnings in the Fall of 1988, at the conclusion of the 1992-93 academic year it had involved over 4,800 students and 34 instructors (17 faculty and 17 senior graduate assistants) over the 5-year period. Most importantly, it has been institutionalized into the Clemson curriculum with plans to substantially expand. During the 1992-93 year, for example, pilot classes in Advanced Engineering Mathematics were taught using the supercalculators. Impetus from this program is also spilling over into other areas of instruction; by the Fall of 1994, Clemson plans to fully integrate graphics calculators into its college algebra and precalculus courses (31 class sections each year with over 1,000 enrollments) and its business calculus courses (another 55 class sections with over 2,100 enrollments).

Background and Origins

The project began in the 1987-88 academic year when a half-dozen members of Clemson's faculty, recognizing the potential of the newly-released HP-28C calculator (January 1987) to help revitalize instruction and learning in basic undergraduate mathematics courses, committed themselves to the task. Because of the high degree of student enthusiasm and response in an early pilot course in calculus taught in 1987-88 by colleague John Kenelly, the project applied to the FIPSE program for funding.

Clemson's Department of Mathematical Sciences felt qualified to integrate new technology, having a history of successful curriculum innovations. Its faculty is more than just mathematicians -- it consists of statisticians, operations research analysts, mathematics educators and computing scientists working together to present integrated programs.

The design of the project was comprehensive and broad-based. Specifically, it sought to design, test, and implement calculator-based courses in single-variable calculus (two courses), multivariable calculus, differential equations, linear algebra, and statistics. These six courses are the mainstream service courses in mathematics for students in science and engineering. The design called for experimental "prototype" and "pilot" offerings of each of the targeted courses during the first two years, followed by a large scale introduction into Clemson's program during the third year. Because of the unusual scope of the project, Hewlett Packard loaned 95 of the HP-28S units to get the project started, and the local administration purchased another 55 HP units and 30 Sharp EL-5200 units. During the prototype and pilot phases, the 150 HP units were used in the five courses in calculus, differential equations and linear algebra, while the 30 sharp calculators were used in statistics classes.

Project Description

During the 1988-89 school year the project designed and class tested prototype versions of each of the 6 courses. Other than some very tentative material from Hewlett Packard, nothing was available relative to the pedagogical use of the calculators; thus the prototype year was genuinely a bootstrapping effort. Part of the summer of 1989 was spent refining the prototype versions, and pilot courses were taught each semester during the 1989-90 year.

During the first two years, class sizes were held to 30 and the calculators were loaned to students each semester for the duration of their courses under signed Calculator Loan Agreements. One senior graduate student acted as a calculator resource person to support the multivariable calculus and differential equations courses. Each of the prototype and pilot courses addressed the following questions.
1. Where is calculator use appropriate--or inappropriate--and why?
2. What does calculator use cost in terms of time and distraction from standard material?
3. Which topics can be more efficiently studied with calculators?
4. What is a proper balance between calculator use and hand performance using traditional methods?
5. Can calculators genuinely enhance conceptual understanding?
6. Which, possibly new, topics can be introduced because of the freedom provided by the calculators?

The standard departmental-wide texts and syllabi were used. But regular, often daily, use of the calculators for both classwork and homework enabled instructors to concentrate on graphical aspects, encourage exploration and experimentation, require active, in-class participation, and provide interesting and realistic approaches. However, the project was not interested in adding substantial amounts of new material, requiring significant calculator expertise, or using high-level calculator routines that delivered "final answers". The primary focus was in increasing student interest, involvement, comprehension and retention of the course material.

During the 1989 Fall semester the project director learned that Hewlett Packard would release its new "supercalculator" in the Spring of 1990, the HP-48SX. This calculator has many important features: excellent integration of graphics, numerics and symbolics, easy memory management, versatile menu-driven software, 32 Kbytes RAM expandable to 288 Kbytes, serial interfacing with PC's and Macintosh's, calculator-to-calculator infrared interfacing, and high-level, plug-in application cards. When combined with the inherent portability, these features made a strong and convincing case for using the new calculator in the future. The project thus advertised that the calculator-enhanced courses for 1990-91 would require either an HP-28S ($150) or an HP-48SX ($250). Even at $250, the 48SX was a bargain when compared in constant dollars to the $25 slide rule of twenty-five years ago that was required in many science and engineering courses.

After a summer of refinement of the course material, the project began its third year, 1990-91, by filling 17 class sections of calculator-enhanced courses during the Fall semester, the more expensive HP-48SX units outnumbering the older HP-28S models by 3-to-1. Campus enthusiasm was high and another 16 sections filled during the Spring term. Preliminary editions of five course supplement manuals were published (Calculus I and II were combined into a single manual), and for the first time students had more than photocopied handouts to supplement their texts.

By Fall 1991 the use of the HP-28S was abandoned altogether because of the availability of the HP-48S, which is functionally the same as the 48SX, except that the 48S cannot accept plug-in memory expansion cards. Except for the engineering student who wishes to insert a high-level application card (e.g., a mechanical engineering card), or the chemistry student who wishes to insert a chemistry card (complete with periodic table in spreadsheet format and capable of molecular weight calculations), the 48S is perfectly adequate. If the 32K RAM of user memory fills up, it is simple to transfer any unused data and programs to a microcomputer disk by serial transmission with a PC or Macintosh. This can later be downloaded to the calculator if needed.

During 1991-92, Clemson taught 52 sections of its calculator-enhanced courses and in 1992-93 another 58 sections. Standard texts are used for both calculator-enhanced and non-calculator versions of the courses, but the pervasive use of technology on the campus has led to the adoption of new texts in almost all of the courses. The basic texts are supplemented with course supplement manuals. The faculty has failed to respond to the project's attempt to support large scale offerings of a calculator-enhanced course in statistics, and the calculator-enhanced version was dropped in 1992-93.

Impact on Students

The impact of the supercalculators has been noticeable and overwhelmingly positive. When all students are equipped with their own calculators, they use them almost daily, feedback is immediate, and a strong element of participation and interaction is evident.

A new dynamic has been introduced into the classroom and the learning process. In calculus, the real benefit of the HP's has been to encourage students to learn--and faculty to teach--the concepts and methods in a more active, constructive environment from analytical, graphical and numerical perspectives. The effective integration of the graphical and numerical solve features of the 48SX have proven to be especially beneficial in helping students to establish visual and numerical connections to the analytic presentations which are characteristic of most textual material. With the calculators, students generally seem to be more involved with the material and they often exhibit a strong sense of personal ownership of their results. And students see them as particularly applicable to their needs for they work equally well everywhere.

In the differential equations and linear algebra courses the 48SX units have dramatically changed instruction. In differential equations they are used almost daily to produce graphs of solutions and to do numerical computations that are impractical by hand. Graphs of solutions especially enhance students' understanding by directing attention to such characteristics as asymptotic behavior and sensitivity to changes in parameters. Solutions to alternative models are also compared graphically and then used as a basis for discussion on the appropriateness of the models. As a computational device, the calculators enable students to identify appropriate problem parameters from observed data and to study the limiting behavior of both discrete and dynamical systems. The rapid determination of eigenvalues and eigenvectors and the solution to systems of linear differential equations permits a much deeper study of such systems in the limited time available.

Supercalculators enhance linear algebra primarily by removing the computational burden associated with hand performance of matrix algorithms, allowing beginning students to focus more clearly on the underlying concepts and theory. We are careful not to use programs that present results at the expense of the students' becoming involved with the underlying mathematical processes. Generally, the programs are interactive and require input and control at key steps. Thus, for example, our Gaussian elimination routine requires that the students decide when and where to pivot, and which row interchanges are needed. And although we are not concerned with introducing a substantial amount of new material into the course, the calculators have enabled us to achieve good results with two modern topics that were often omitted in previous offerings: the interpretation of Gaussian elimination as an LU-factorization and its application to linear systems with multiple right-hand sides, and the interpretation of the Gram-Schmidt process as a QR-factorization and its application to least squares problems. These topics are important today because they lie close to the heart of many computer codes used to handle large linear systems.

More than anything else, regular use of the calculators throughout our courses *has changed not only what and how we teach, but also what and how we test.* We allow free use of the devices on our tests, and part of the learning process is to determine when, and when not, to use them. There is plenty of room for both theoretical and computational questioning, and we have been unable to obtain this level of testing in a more computationally restricted environment.

As the HP-48S/SX units have been made available on campus through the impetus of this project, their use has spread rapidly to other units such as civil engineering, electrical engineering, chemistry and physics. Though no data are available to document this impact, students in formal interview sessions relate many examples of using their calculators in other classes, of instructors structuring lectures and demonstrations with them, and of uses outside of class. Many students are maintaining extensive libraries on their calculator units. Although the first systematic use of the supercalculators was in mathematical sciences, it is spreading to other areas of the university.

Do students learn more mathematics? Do they better understand what they learn? These are tough questions, questions that we have been unable to answer with any strong sense of accuracy. But our students have certainly seen mathematics in a different light, have clearly shown us that they can grasp some of the concepts better than before, and are all more interested and involved in their learning. Overall, we see this as a positive effect.

Impact on Faculty

To incorporate supercalculators effectively into undergraduate mathematics it became apparent that the faculty involved must do two things. First, change their role as instructor from being a traditional lecturer who transmits knowledge to students to being more of a professional guide who may lecture at various times but most often explains, questions, challenges and in a variety of other ways helps students become *engaged* with the course material. Second, relinquish their hold on the traditional controlled college classroom atmosphere, characterized by "teachers talking - students writing", in favor of a more unstructured environment in which students themselves become the more active participants.

In short, our time-honored method of lecturing to our classes has given way to other, more constructive arenas for learning. This has often been a painful experience because *change in the classroom does not come easily to most mathematics faculty*. The thought of teaching mathematics in a more unstructured setting wherein the instructor is to, somehow, "engage" students is foreign and, indeed, threatening to some. It means, for instance, that college faculty can no longer be the authoritarian purveyors of knowledge and must rethink, very carefully, not only what they will do but how and why it should be done. In almost every case (but not all), the faculty who became involved in this project were able to make the necessary adjustments. In so doing, they experienced what much current research tells us about how children learn mathematics. People learn best when they construct their own personal interpretations of concepts by making the necessary connections; they learn little in the passive process of "being told", because they fail to establish the connections that ultimately lead to knowledge and understanding as something personal.

The project has had an unexpected impact on faculty; it has generated a renewed interest in, and enthusiasm for, good teaching and the pedagogical issues surrounding the good teaching of the affected courses. During the 1990-91 academic year, the project conducted a weekly support seminar for the 15 instructors who were teaching in the project. That seminar proved to be a lively and often spirited forum for the exchange of ideas, opinions, student activities and projects, what was working well in the classroom and what was not. Most significantly, it was the *first* such seminar devoted to teaching issues conducted by the department in many years. It was expanded considerably during the 1991-92 academic year for the 22 instructors involved in the project by conducting three seminars each week, one for each of the single-variable calculus, multivariable calculus, and differential equations courses.

The most powerful result may be the faculty's observations about how student learning is being fundamentally changed by the supercalculators. That is, they have observed an empowerment of the students to use the calculators to *construct highly personal interpretations of the mathematics*. In particular, students develop strategies to generate, manipulate, and use visual images relevant to their mathematics. In the past, visualizations have been largely a *product* of students' mathematics rather than a *process* used to develop mathematical ideas. There is need for research on this topic and the Clemson environment may be one where a useful and productive research program could be mounted.

Student Perceptions

We have no objective measures of the student learning that has taken place under the project, that is, traditional assessments such as performances on common exams given to students involved in the project vis-a-vis a control group. It is practically impossible to match student groups in terms of overall mathematical backgrounds, abilities, skill levels, attitudes, and learning styles. During the prototype and pilot phases, students who participated in the project were those who, for whatever reason, happened to be enrolled in the class sections targeted for calculator enhancement and these sections were not identified ahead of time as being anything special. Since then, the students who participate are those who consciously choose to enroll in class sections that are advertised as being calculator-enhanced classes that require the purchase of an appropriate calculator. Because of this, it is not proper to compare performances on common exams. It would not be fair to those taught in the traditional way, nor to those in calculator-enhanced classes. The learning environments and expectations are substantially different.

However, during the development of the project an external evaluator obtained feedback of student perceptions from 61 classes and 1,523 students. The questionnaire that was used to assess the students' perceptions of the use of supercalculators contained 17 statements that called for responses on one of five levels, from strongly agree to strongly disagree. Each of the statements dealt with some aspect of using calculators in the courses. There were also questions asking for open-ended responses to "What did you like best?", "What did you like least?", and "Give an example of a problem that you have solved using the calculator that someone without a calculator probably could not do." We present representative summaries of the results on six questions selected from the questionnaire.

We are interested in students' perceptions about the role of supercalculators in helping them *understand*, as opposed to the more obvious role of routine calculation. Table 1 shows selected statements which illustrate students' perceptions, with the percent of students' responses for strongly agree (SA), agree (A), neutral (N), disagree (D), and strongly disagree (SD). These data were gathered from students at the end of each course.

For statement 1 approximately 74 percent of the students across the courses felt the supercalculators helped them understand the material. A similar question for the Calculus I course with data from 9 classes involving 268 students produced even higher percentages, so the positive perceptions were as likely from beginning students as they were from more advanced students. Follow-up responses, both written and oral, provided extensive documentation of the reasons students felt the calculators helped them understand the material. Among the more frequent categories of statements were being able to obtain graphs quickly and accurately, trying several graphs, reasoning from graphs; ease of computations, accuracy of computations, matrix computations; visualization (you could "see" mathematical relations).

Table 1. *Students' Perceptions About the Use of Calculators*

Item	Percent				
	SA	A	N	D	SD
1. The graphics calculator helped me understand the material in the course.	23	51	16	7	2
2. The graphics calculator allowed me to do more exploration and investigation in solving problems	31	44	19	5	2
3. Learning the calculator was so difficult that it detracted from learning the material in the course.	2	6	14	45	33
4. Time devoted to instruction in the use of the graphics calculator meant less material was covered in the course.	3	9	16	47	25
5. The graphics calculator helps me have better intuition about the material.	16	45	28	9	3
6. I would recommend that entering freshmen seek out courses using the graphics calculator.	29	42	21	5	3

Over 91 percent of the students agreed (SA or A) with a statement that the calculators were *useful* in solving problems in these courses. In many classes 100% agreed or strongly agreed. It was clear that they were responding to ease and accuracy of calculations in doing the problems as well as being able to approach more difficult or more realistic problems. A considerably smaller percentage of students, but still a majority, felt the calculator was *necessary* in solving problems. The ability to store programs and recall them for working on complex problems was mentioned often by students as evidence of the calculator's capability as a problem solving tool.

Exploration and investigation are desirable aspects of a dynamic, process-oriented approach to mathematics and the project felt that using supercalculators would facilitate these traits. Statement 2 (Table 1) was designed to determine if students sensed an opportunity for exploration and investigation in calculator-enhanced courses. Over 74 percent of the students across all classes sensed such an opportunity. These are not characteristics generally associated with the study of lower division undergraduate mathematics. The results are even more notable in that another 19 percent of the students were neutral and only about 7 percent explicitly disagreed with describing their calculator-based mathematics classes to have more exploration and investigation. Comments throughout the open-ended responses and in the interviews underscored this result.

Some people have perceptions that using supercalculators in these courses will be detrimental because of attention or time given to the calculator rather than to the substance of the course. Two items (statements 3 and 4 in Table 1) dealt with these issues. A substantial majority of students disagreed with these statements, that is, they did not feel that using the calculators short changed them in terms of their course content. Many of the open-ended responses pointedly addressed that the calculator-based courses demanded extra effort and extra time. In interviews, a frequent theme was hope for some way to ease the strain of learning to use the calculators (e.g., special seminars, graduate assistants, handouts). Some students were concerned about the coverage of material before taking the courses; but at the end, almost all of them felt they had covered *more* material than friends in non-calculator courses.

Intuition about mathematics is rather intangible but related to a feeling of understanding and having self-confidence in doing mathematics. Statement 5 (Table 1) examined the students' perception of their mathematical intuition. Here again, over 60 percent of the students would agree or strongly agree with the statement. In student interviews, those who were successful with the calculator (and almost all were) were very confident in their knowledge of mathematics.

Finally, students responded to a question about their recommendations to other students (statement 6 in Table 1). This is clear endorsement of the calculator-enhanced program by the people most able to take issue with it - the students at the end of their courses.

Summary and Conclusions

Not only has Clemson shown that high-level supercalculators can be effectively integrated into the undergraduate mathematics curriculum in a comprehensive way, but that they also offer special advantages not readily available with mainframe or microcomputing equipment. Advantages such as their portability, modest cost, their unexpected ability to help students become engaged with mathematics on a personal level and, above all, their role in changing the testing environment. With technology, it is mandatory that we change our tests, so that they more accurately reflect the knowledge and skills that we want our students to acquire. *Changing testing is easy and natural with supercalculators,* but difficult to do with microcomputers.

The project has been immensely successful in achieving large-scale implementation of its results on the Clemson campus. While others are struggling to get one or two class sections going, Clemson's more than 50 class sections are moving forward rapidly. And in that process, there is being generated a new enthusiasm for the teaching, and a new approach to the learning, of mathematics by the faculty and the students who are involved. Student reactions to calculator use are well documented in the report by the external evaluator, and they are overwhelmingly positive. We have neglected to systematically ascertain changing faculty attitudes, but more than a few have commented that they cannot conceive of teaching *without* supercalculators again. Our project is clearly institutionalized and can only get better as it matures. We are convinced that the impetus for change brought forth by this project is both strong and long-lasting. We will look back a decade from now with a genuine understanding and a clear realization of what it is helping to accomplish.

Acknowledgment

Three-year funding for the project was provided by the U.S. Department of Education's FIPSE program (1988-1991).

Teaching Mathematics with Calculators (TMC): A Local and National Inservice Teacher Education Project

John G. Harvey
University of Wisconsin - Madison

In 1989, professional mathematics and mathematics education organizations began to publish reports that establish new standards for mathematics education at the school, undergraduate, and graduate levels and for the mathematical and professional preparation of preservice and inservice teachers at all grade levels (Leitzel, 1991; National Council of Teachers of Mathematics [NCTM], 1989, 1991; National Research Council [NRC], 1989, 1990, 1991). These new standards for mathematics education are being established in response to the general criticisms of schools and schooling (e.g., National Commission on Excellence in Education, 1983), as a continuing effort to improve mathematics instruction (NCTM, 1980), in recognition of the poor performance of U. S. students on national and international measures of achievement (Dossey, Mullis, Lindquist & Chambers, 1988; Lindquist, 1989; McKnight, Crosswhite, Dossey, Kifer, Swafford, Travers, & Cooney, 1987), and to respond to changing conditions for and outdated assumptions about mathematics (e.g., NRC, 1990). Each of the reports seemed to be based upon a transition to the future that, to me, is summarized well in *Reshaping School Mathematics: A Philosophy and Framework for Curriculum* (NRC, 1990, p. 5):

> Evidence is mounting from many sources that our present curriculum must change course if it is to serve society well in the twenty-first century. Forces for change, which are growing increasingly powerful, are beginning to redirect the mathematics curriculum in several important ways:
>
> The focus of school mathematics is shifting from a dualistic mission--minimal mathematics for the majority, advanced mathematics for a few--to a singular focus on a significant common core of mathematics for all students.
>
> The teaching of mathematics is shifting from an authoritarian model based on "transmission of knowledge" to a student-centered practice featuring "stimulation of learning."
>
> Public attitudes about mathematics are shifting from indifference and hostility to recognition of the important role that mathematics plays in today's society.
>
> The teaching of mathematics is shifting from preoccupation with inculcating routine skills to developing broad-based mathematical power.

The teaching of mathematics is shifting from emphasis on tools for future courses to greater emphasis on topics that are relevant to students' present and future needs.

The teaching of mathematics is shifting from primary emphasis on paper-and-pencil calculations to full use of calculators and computers.

Of greatest interest here is the last of these points; that is, that mathematics teaching and instruction must shift to the "full use of calculators and computers." Each set of standards that has been published affirms this position; the amount of emphasis on this point ranges from a single statement (NCTM, 1991) to total integration of the use of technologies into both the curriculum and classroom instruction (Leitzel, 1991; NRC, 1989, 1990, 1991). For example, *Reshaping School Mathematics* returns to this theme on at least five other occasions (pp. 8, 16, 20-21, 37, 43). This universal endorsement of the integrated uses of calculator and computer technologies creates a challenging goal; we must now find ways to achieve this goal by (a) developing appropriate curriculum materials and assessments, (b) changing mathematics instruction and the knowledge and skills of our students, and (c) providing preservice and inservice teacher education on the integrated uses of technologies. This paper describes a project that has as its goals the inservice education of teachers at local and national levels so that they will be better prepared to integrate the uses of calculators into their day-to-day mathematics instruction.

Technologies: A Teacher Education Problem

When I first began to specialize in mathematics education, the nation was in the "throes" of the "New Math" movement. That movement has been, in recent times, the only attempt to reform the mathematics curriculum that is roughly of the same size and scope as is the effort recently mounted to implement the newly announced standards for mathematics education (the New Standards). As a result, this is the only mathematics education reform movement to which we can look for clues about the ways in which we should proceed in order to implement the New Standards.

The New Math movement did not achieve its goals, but it was not a failure either. When I look at schools, teachers, and students of today I see continuing, positive effects of the New Math movement; examples of these effects are (a) the improved mathematics preparation of secondary school teachers, (b) the Advanced Placement Program in mathematics, (c) an elementary school mathematics curriculum that is no longer solely focused on arithmetic, and (d) the increased mathematics preparation of students entering college.

The present failures of mathematics education should not be laid entirely at the door of the New Math movement. Many of the things decried in today's reports concerning the failure of our schools and our mathematics curriculum are holdovers from before the "New Math;" these holdovers include (a) an emphasis on drill-and-practice, (b) teacher-centered instruction, (c) problems whose solutions depend almost entirely upon low-level skills, and (d) a narrow view of mathematics as a collection of techniques, algorithms, and tricks.

Since the New Math movement was not a failure and not the origin of some of the problems connected with the present mathematics curriculum, what lessons are to be learned from the attempts to install New Math in our schools--and colleges?
1. One lesson we have learned is that teachers when convinced that change may lead to improvement will eagerly embrace change or, at least, try to and seem to embrace change. Thus, if we can convince teachers that the New Standards will produce positive changes, I believe that they will readily embrace these goals.
2. Second, we must be careful to look beyond the surface characteristics of the curriculum materials and instructional techniques developed that purport to implement the New Standards and discover if these materials and techniques are in accord with the philosophy of the New Standards. Too many of the sets of curriculum materials developed for the New Math movement did not really agree with the philosophy of that reform movement.
3. Third, both school and college faculty must have ownership of these new ways of teaching and learning mathematics. Ownership of the new standards means believing in and being willing to implement the goals espoused by them, *but* it also means that teachers must be adequately prepared to teach them. The New Math movement failed to prepare most teachers adequately to implement the new curricula that they were given; most teachers did not already have the necessary mathematics backgrounds or the flexibility needed to take those materials and implement them properly. Implementation of the New Standards will also fail if we do not engage in inservice and preservice teacher education for *all* mathematics teachers. One of the major topics of the preparation that teachers will need in order to own the New Standards is the uses of calculator and computer technologies in mathematics instruction and learning.

Some, possibly many, people do not believe that teachers need to be prepared to use calculators or computers effectively. After all, it only takes a few minutes to learn to use many calculators, and so they argue, why should teachers need extensive inservice training on their use in mathematics instruction? This argument simply is not true because among other things it equates these tools with many of the other simpler instructional tools that we use (e.g., the overhead projector). For example, how many elementary and middle school mathematics teachers have discovered, on their own, that the ability of the Texas Instruments Math Explorer to reduce fractions can be used to find the prime factors of three-digit numbers? How many have discovered that the integer divide function on the Math Explorer can be used to express any number in base 5? base 7? base n? How many high school teachers have discovered that the parametric function mode of the Texas Instruments TI-81 makes it easy to graph the inverse relation associated with a function? How many of them have discovered the elegant way of showing that the *sin* and *cos* functions are graphs that result from "unwrapping" the unit circle? So, while some of the calculators may be easy to learn to use, if we expect their integration into mathematics instruction as envisioned by the *Curriculum and Evaluation Standards for School Mathematics* (NCTM, 1989) and *A Call for Change: Recommendations for the Mathematical Preparation of Teachers of Mathematics* (Leitzel, 1991), we must find a way to provide sufficient inservice on how one "does" mathematics using these tools, how one teaches students to use these tools effectively, and how one integrates the use of these tools into mathematics instruction. *Teaching Mathematics with*

Calculators: A National Workshop is a project that is attempting to provide materials and methods that will help to solve this problem.

Teaching Mathematics with Calculators: A National Workshop

Teaching Mathematics with Calculators: A National Workshop (*TMC*) is a joint project of the Mathematical Association of America and the National Council of Teachers of Mathematics; it is funded by the National Science Foundation and Texas Instruments Incorporated. The goals of the *TMC* Project are these:
1. to assist middle and high school teachers in two demonstration school districts to integrate the effective use of calculators into their mathematics instruction,
2. to develop instructional packages that will assist mathematics teachers to integrate the effective use of calculators into their instruction effectively, and
3. to disseminate the instructional packages as broadly as possible so as to create *The TMC National Workshop*.

The key to achievement of all three of these goals is attainment of the first one; that goal is being attained through a series of *TMC* Summer Institutes.

The Demonstration School Districts

The two *TMC* demonstration school districts are the Mesquite (TX) Independent School District and the Ft. Worth (TX) Independent School District. Though both are in the Dallas-Ft. Worth Metroplex area, they are very different from each other. Mesquite abuts the eastern city limit of Dallas and was largely a rural school district until the population of Dallas and its surrounding towns grew after World War II; Mesquite can probably be accurately termed a "bedroom community." The Mesquite district has experienced a large increase in its student population in recent years. At present the Mesquite school district has about 25,000 students in its elementary, middle, and high schools; about 20% of these students are black, Hispanic, Asian, or American Indian. Ft. Worth is smaller than Dallas but is considerably larger than Mesquite. The Ft. Worth school district enrolls about 65,000 students in its elementary, middle and high schools; the school population is almost equally divided between black, Hispanic, and white students. The *TMC* Project sought the participation of all of the middle and high school teachers in the Mesquite school district and the middle and high school teachers in three of the "pyramids" in the Ft. Worth district. The numbers of students and teachers in the Ft. Worth pyramids are approximately the same as in the Mesquite district.

TMC Teacher Education Activities

The model for teacher inservice education being used by the *TMC* Project comes, in part, from a successful University of Wisconsin-Madison Public Schools Cooperative College-School (CCSS) Project that started in 1969 and lasted into the early 1970's. The CCSS Project staff, with the help and advice of the building principals, selected one teacher in each school to be a member of the CCSS Teacher Cadre; this teacher was selected based upon his or her willingness to participate and his or her potential to be a lead teacher is his or her school. The members of the CCSS Teacher Cadre were participants in an initial summer institute program on the University of Wisconsin (UW) campus. Throughout the remainder of the project Cadre members were released one day each week from their usual teaching duties. During half of this day they worked with the other teachers in their building to improve mathematics instruction in that school; during the other half day each week these teachers convened in a seminar led by UW faculty. The members of the CCSS Teacher Cadre worked very effectively with the teachers in their buildings, but as a group they were also very effective in influencing mathematics education across the district. For example, as long as the cadre existed it helped to select the mathematics textbooks used throughout the district and helped to provide inservice training for teachers in the district.

Emulating the CCSS Teacher Cadre model, the *TMC* Project has selected middle and high school teachers to be Mathematics Technology Lead (MTL) Teachers. In selecting the MTL Teachers the project was careful to select at least one mathematics teacher from each building and to select teachers that would be able to work effectively with the other mathematics teachers in their school. The MTL Teachers have received or will receive more extensive inservice education from the project faculty than do their colleagues though both sets of teachers are invited to attend *TMC* Summer Institutes. All of the teachers who participate in the project receive instruction on the use and integration of several different calculators including four-function, scientific, fractions, and graphing calculators. All of these calculators are included because schools have limited ability to purchase calculators and must use what is available to them.

The 1990 TMC Summer Institute

The first summer institute for teachers was held during the summer of 1990; 21 Mesquite and three Ft. Worth middle and high school teachers participated in the 1990 *TMC* Summer Institute; there were 20 female and four male participants. The Ft. Worth teachers served as liaisons from their district to the Mesquite district. The teaching experience of the participants ranged from a single year to 20 or more years. The Mesquite school district lacks funds to support summer programs for teachers; aside from the inservice days that precede the beginning of school each year, there are no teacher inservice days in this district. Thus, the Mesquite MTL Teachers had no experience in developing curriculum materials and had had little opportunity to work with each other except on an informal basis.

The 1990 Summer Institute was held at the North Mesquite High School; the Project Co-Directors, John Kenelly and I, were the resident faculty members. As the resident faculty members, we were responsible for planning for and providing a majority of the instruction on calculator use and instructional integration during the two, 75-minute morning sessions. During the three-hour afternoon curriculum workshop sessions, the Co-Directors worked closely with the Institute participants as they, in groups, prepared initial drafts of the printed materials that would accompany videotapes on graphing calculators and the fractions calculator.

In the morning workshop sessions there was instruction on the abilities of the five calculators and their uses as mathematics tools; the instructors were the resident and visiting faculty members. The visiting faculty were college faculty who are leaders in calculator-based mathematics instruction. In these sessions, all of the functionalities of each of the calculators were discussed; the matrix, statistics, programming, and parametric function functionalities of the Texas Instruments TI-81 were introduced but not fully explored.

At the 1990 Summer Institute each of the participants was given a personal set of calculators that included all of the calculators studied except for the Texas Instruments TI-81. Each participant received a TI-81, but it remained the property of their school district. In addition, Texas Instruments gave the Mesquite school district enough Math Explorer and TI-81 calculators so that there are one or more classroom sets of fractions calculators in each middle school and one or more sets of TI-81 calculators in each high school. We urged the MTL Teachers to use these calculators and the materials they had developed during the upcoming school year.

1991 *TMC* Mesquite ISD Summer Institute Programs

During the summer of 1991, two *TMC* Summer Institutes were held in the Mesquite school district. The first one, the 1991 Mesquite MTL Teacher *TMC* Summer Institute, was a four-week institute for the MTL Teachers trained during the summer of 1990; this institute program will be described before discussing the second Mesquite TMC Summer Institute.

1991 Mesquite MTL Teacher TMC Summer Institute. Twenty MTL Teachers (17 females; 3 males) attended this institute; 18 of these participants were participants in the 1990 *TMC* Summer Institute while two were new MTL Teachers. Six of the 24 teachers who participated in 1990 did not participate in 1991; of these two had left teaching, one had become an assistant principal, one had moved to another school district, one was teaching summer school, and one elected not to participate in 1991. Thus, there was only one true dropout from among the MTL Teacher cadre from the summer of 1990 to the summer of 1991. The 1991 Mesquite MTL Teacher Summer Institute was divided into two, 2-week parts.

During the first two weeks of this institute the participants spent mornings with the Project Co-Directors discussing the functionalities and uses of the calculators that are being used by the Project: the TI-81, the Math Explorer, the TI-34, the TI-30, a TI business calculator, a "chain logic" four-function calculator, and an "algebraic logic" four-function calculator. During these classes questions about the calculators that the teachers had framed during the 1990-91 school year were answered, and the more advanced features of the calculators were explored. In particular, we spent a considerable amount of time discussing the numeric features of the advanced scientific calculators (i.e., the TI-81 and the TI-34), the matrix, statistics, programming, and advanced graphics capabilities of the TI-81, the differences between the two four-function calculators, and the problems that can be solved--and the ways that they are solved--using a business calculator. In addition, we discussed with the MTL Teachers the possible vignettes for three new *TMC* videotapes--one about the graphing calculator, one about the fractions calculator, and one about four-function calculators.

For the first part of this Institute the afternoons were spent by the teacher participants in developing (a) mathematics curriculum materials for the middle and high school grades that integrate the use of calculators into instruction and learning and (b) plans for the instruction that they would provide as part of the *TMC* institute given for the Mesquite middle and high school teachers not yet trained. Each of these activities is briefly described next.

During the summer of 1990, the MTL Teachers developed printed materials intended to accompany videotapes on the uses of graphing calculators and of fractions calculators. Using the experience gained during the 1990-91 school year, the first curriculum development activities of teachers in the 1991 Mesquite MTL Teacher Summer Institute were the editing and refinement of the materials they had developed in 1990. In order to edit and refine these materials they divided themselves into (a) a group consisting of all of the middle school teachers and (b) two groups of high school teachers of approximately equal size. When these three groups finished revising the 1990 materials they turned to the development of new materials. The middle school teacher group developed additional materials for the Math Explorer, for four-function calculators, and for scientific (non-graphing) calculators. One of the two high school teacher groups spent a considerable amount of time exploring ways in which graphing calculators could be used in geometry instruction; among other things they developed a way of using the TI-81 as an electronic geoboard and developed geoboard activities using this application. The other high school teacher group developed additional, advanced activities for graphing calculators. None of the groups developed activities using the matrix, statistical, or programming capabilities of the TI-81; these capabilities will be further explored and activities developed using them during the 1992 *TMC* Summer Institutes.

In 1990 only a very few of the MTL Teachers had experience in developing curriculum materials. As a result the amount and the quality of the material that was developed was not as good as we had expected. During 1991 these same teachers showed considerable maturity in the development of curriculum materials, in their knowledge of calculator functionality, and in the uses of calculators in mathematics instruction. Thus, the materials revised and refined during 1991 and the new materials developed during this summer by these same teachers were of high quality.

In addition to revising, refining, and creating curriculum materials during the afternoons of the first part of this Summer Institute, the teacher participants also developed plans and materials for the *TMC* Summer Institute for Mesquite Middle and High School Teachers convened for those Mesquite middle and high school teachers who are not MTL Teachers. The planning of the MTL Teachers centered around ways in which they might work with their colleagues to show them places and ways of integrating calculator use into their present curriculum materials and syllabi.

The second part of the 1991 Mesquite Summer Institute for MTL Teachers was concurrent with the 1991 *TMC* Summer Institute for Mesquite Middle and High School Teachers. During the second part of their institute program, the MTL Teachers spent their mornings continuing to develop curriculum materials and to refine and revise their plans for working with their colleagues during the afternoons of this two-week period. They met jointly with the participants in the other institute on the two occasions when there were visiting faculty. The Mesquite MTL Teachers spent their afternoons during this part working with their colleagues in both small and large groups to discuss and to develop ways of integrating calculator use into Mesquite classrooms using their present textbooks and syllabi.

1991 TMC Summer Institute for Mesquite middle and high school teachers. This two-week Summer Institute was for middle and high school Mesquite teachers who are not members of the Mesquite MTL Teacher cadre.

All of the middle and high school teachers who are not Mesquite MTL Teachers were invited to participate in this Institute; there were approximately 57 teachers in this group. Thirty-two teachers in this group accepted the invitation to participate in this institute on a full- or part-time basis. Those teachers who attended on a part-time basis were ones who were engaged in other activities (e.g., summer school teaching) from which they were unable to obtain release. We decided that we would permit these teachers to attend on a part-time basis since we sought to train as many of the Mesquite middle and high school teachers as possible.

The morning program for this *TMC* institute consisted of workshop sessions in which the Co-Directors worked with the participants to help them learn about the calculators and ways in which they might be used effectively in mathematics instruction. During each of these workshop sessions, small groups of MTL Teachers were present to serve as teaching assistants or "help buttons." On four days visiting *TMC* faculty members also pursued these themes with the participants.

During the afternoon sessions, the participants in this Summer Institute were taught by the Mesquite MTL Teachers. In general, during these sessions the participants met in groups consisting of high school or middle school teachers who taught the same subjects or the same grades. The syllabi and textbooks for these subjects and grades were carefully examined in the afternoon sessions and ways of integrating calculator use into these materials were explained and developed.

1991 Ft. Worth *TMC* Summer Institute Program

Only one summer institute for Ft. Worth teachers was held in 1991; this was a two-week Ft. Worth *TMC* Summer Institute for MTL Teachers. Twenty-nine Ft. Worth middle and high school teachers (21 females; 8 males) attended this institute; one of the participants was a Mesquite MTL Teacher. As in the Mesquite Independent School District the age and experience of the teachers varied considerably.

The teacher participants in the Ft. Worth Summer Institute were much more experienced in the use of calculators in mathematics instruction and in developing curriculum than were the Mesquite teachers who attended the 1990 *TMC* Summer Institute for MTL Teachers. For these reasons the program of the 1991 Ft. Worth Summer Institute was faster paced and as a result, covered all of the content of the 1990 Mesquite Summer Institute and portions of the material covered in the 1991 Mesquite MTL *TMC* Summer Institute. As in the other two 1991 summer institutes, during the mornings, the Co-Directors or a *TMC* visiting faculty member discussed the capabilities of several different calculators with the participants and ways of using those calculators in mathematics instruction.

The teacher participants in the 1991 Ft. Worth institute developed materials for fractions and graphing calculators during the afternoon sessions. While developing these materials the teachers were in groups of four or five each. The afternoon sessions were, in general, managed by Bev Ramsey, Gwen Kann, and Karen Flowers. Ms. Ramsey and Ms. Kann are the Ft. Worth teachers who were trained as MTL Teachers in the 1990 *TMC* Summer Institute for MTL Teachers and the 1991 Mesquite MTL Teacher *TMC* Summer Institute; Ms. Flowers is a Mesquite MTL Teacher.

1992 *TMC* Summer Institutes

During the summer of 1992, three *TMC* Summer Institutes are scheduled. One of these institutes will convene in Mesquite for the MTL Teachers trained during the summer of 1990. During this two-week institute the MTL teachers will (a) edit and refine the printed materials that they developed during the summer of 1991 and (b) discuss with the Co-Directors the themes of the remaining videotapes and instructional packages that will be developed.

The second and third 1992 *TMC* Summer Institutes will convene in Ft. Worth. One of these institutes will meet for three weeks and will be for the Ft. Worth MTL Teachers who participated in the 1991 *TMC* Summer Institute for Ft. Worth MTL Teachers; this institute will be like that one conducted for MTL Teachers in Mesquite in the summer of 1991. The final *TMC* Summer Institute conducted in 1992 will be one for middle and high school teachers from the three Ft. Worth pyramids that are participating in the project; this institute will be like the one conducted for Mesquite middle and high school teachers in 1991.

TMC Instructional Packages

The *TMC* Project is developing instructional packages that will include one or more videotapes and a set of printed materials designed to accompany the videotape(s) in that package. The intended audience of each instructional package is middle and high school mathematics teachers. Each videotape and its accompanying printed materials will be a complete, self-contained package that a teacher or a group of teachers can easily use and, we hope, find worthwhile. To use any one of the instructional packages, a teacher or group of teachers might begin by viewing the problem vignettes one at a time. At the conclusion of each problem vignette, the teacher(s) would turn to that portion of the printed materials that describes both the mathematical content of the problem and the way in which the calculator is used to help solve the problem. Having studied the problem solution, the teacher(s) might rewind the videotape and view the vignette again until both the mathematical content being taught and the way in which the calculator is used as a tool to solve the problem are understood. When each problem vignette has been viewed and studied, the teacher(s) would study the suggested instructional activities in the printed materials so as to develop instructional units or lessons of their own that integrate the calculator.

The *TMC* Videotapes

Only one calculator is used in each videotape. Using that calculator Mesquite and Ft. Worth middle or high school teachers teach classes of students the problem vignettes included in the tape. For the first two videotapes, the problem vignettes were completely specified; that is, a script was developed by the project scriptwriter that specified both the actions and the words of both the teacher and his or her students. This approach was chosen to ensure that the "mathematical correctness" of the problem vignettes, but the resulting videotapes lacked spontaneity--as might be predicted, especially in hindsight. As a result, the technique for developing the videotapes has been changed. The new technique being used can be outlined in this way:

1. The *TMC* Advisory Board proposes the problems that might be taught. (Through June 1992, the *TMC* Advisory Board consisted of the following members: Gary Bitter, John Dossey, Bettye Forte, Shirley Frye, John Jobe,

Lorraine Mitchell, David Pagni, Gay Riley-Pfund, Cathy Seeley, Dorothy Strong, Bert Waits, and Steve Willoughby.)
2. The MTL Teachers discuss and advise the Co-Directors about the proposed problems.
3. Based on the advice of the MTL Teachers, the problems are revised and a final form of them is produced.
4. An outline of each problem is developed that describes the desired mathematical outcomes and the ways that the calculator involved can be used effectively as a tool while the problem is being solved.
5. Once a teacher has agreed to teach the problem vignette, she or he studies the problem outline and prepares a lesson based on that outline; she or he chooses her or his own pedagogical approach. Before the vignette is taught the teacher prepares her or his students by using the calculator with them in ways like the ways in which the calculator will be used while studying the problem in the vignette, but the problem is not taught to the students beforehand.
6. Each vignette is taught three times during videotaping so that teacher, individual students, and class reactions to the problem can be captured.

The result is a videotape in which shows a more comfortable teacher and a more spontaneous response from the class than did the procedure originally used.

Thus far three *TMC* videotapes have been produced, and two instructional packages have been released. The two instructional packages that were released at the April 1992 Annual Meeting of the National Council of Supervisors of Mathematics contained these videotapes:

The Graphing Calculator: Building New Models
The Fractions Calculator: Old Things, New Ways

Six additional videotapes will be developed in 1992, 1993, and 1994. The themes of these videotapes will be:
1. answering parent questions about the uses of calculators in mathematics instruction,
2. ways of using four-function calculators effectively,
3. uses of simple scientific calculators,
4. teaching matrices and discrete mathematics with a "matrix calculator,"
5. using the statistical abilities of advanced calculators to teach quantitative literacy, and
6. programming advanced calculators to enhance their uses as mathematics tools.

The first two of these videotapes will be produced during the 1992-93 school year; the remaining four videotapes will be produced during the 1993-94 school year.

The *TMC* Printed Materials

The majority of the *TMC* printed materials are being developed by the Mesquite and Ft. Worth MTL Teachers during the summer institutes. Thus far, these teachers have developed materials related to the uses of graphing and fractions calculators; many of these materials were included in the first two instructional packages released in April 1992. They have also developed materials related to the uses of four-function calculators and simple scientific calculators. Development of these materials will continue during the summer of 1992 and new materials will be developed for videotape themes (4), (5), and (6).

The materials being developed by the MTL Teachers are teaching suggestions (e.g., lesson plans) that show teachers places in the curriculum where calculators can be effective mathematics tools. It is intended that teachers will use these materials to further their knowledge of the abilities of the calculators and as models of the ways in which calculators can be used effectively. Each teaching suggestion identifies an instructional objective and describes ways to use calculators that achieve that objective.

The remainder of the *TMC* printed materials are developed by the Advisory Board and the Co-Directors. These parts of the printed materials are step-by-step descriptions of the ways in which the problems in the videotape vignettes are solved. So that teachers can follow along while viewing the videotapes, these materials give keystroke instructions.

Dissemination of the *TMC* Instructional Packages

Each of the *TMC* Instructional Packages will be distributed initially at an annual meeting of the National Council of Supervisors of Mathematics by giving to each person attending that meeting a single copy of each package along with a copyright release that permits them to make copies of both the videotapes and the printed materials for instructional purposes. Additional copies of the instructional packages may be obtained from the Mathematical Association of America; the cost of each of these packages will be the cost of duplicating and distributing them.

Conclusion

This paper reports the work and progress of the *TMC* Project at the end of its first two years--two years short of its completion. We have no data that tell us that we have achieved any of our goals. Our efforts to evaluate the effectiveness of our local inservice education program is just beginning. We have only recently distributed the first two instructional packages to a national audience. Thus, I can reach no conclusions based upon reliable qualitative or quantitative data. From anecdotal evidence and my own observations I would conclude the following:

1. We have demonstrated that middle and high school teachers can develop materials that incorporate calculators effectively into mathematics instruction. In order to do this, those teachers must have a carefully designed inservice education program that explores with them the abilities of the calculators to be used and relates those abilities to the content of the mathematics curriculum.
2. As was true for the UW-Madison Public Schools CCSS Project it appears that a cadre of teachers can develop plans for and provide inservice to their colleagues.
3. We have worked most with the MTL Teachers in the Mesquite school district. They are increasingly able to develop calculator-based mathematics curriculum materials even though they had had almost no prior curriculum development experience before. The change in their abilities was exhibited in the summer of 1991 when they took the materials they had developed the previous year and revised them; the improvement in the materials was quite noticeable to me.
4. The project has developed a technique for making videotapes that teaches mathematics and are believable because practicing teachers and their students are the "actors."

Acknowledgment

The project described in this paper is funded by the National Science Foundation and Texas Instruments, Incorporated. The positions and opinions expressed are those of the author and are not endorsed by either funding agency.

References

Dossey, J. A., Mullis, I. V. S., Lindquist, M. M., & Chambers, D. L. (1988). *The mathematics report card: Are we measuring up?* Princeton, NJ: Educational Testing Service.

Leitzel, J. R. C. (Ed.). (1991). *A call for change: Recommendations for the mathematical preparation of teachers of mathematics.* Washington, DC: Committee on the Mathematical Preparation of Teachers of the Mathematical Association of America.

Lindquist, M. M. (Ed). (1989). *Results for the Fourth Mathematics Assessment of the National Assessment of Educational Progress.* Reston, VA: National Council of Teachers of Mathematics.

McKnight, C. C., Crosswhite, F. J., Dossey, J. A., Kifer, E., Swafford, J. O., Travers, K. J. & Cooney, T. J. (1987). *The underachieving curriculum: Assessing U. S. school mathematics from an international perspective.* Champaign, IL: Stipes.

National Commission on Excellence in Education. (1983). *A nation at risk: The imperative for educational reform.* Washington, DC: Author

National Council of Teachers of Mathematics. (1980). *An agenda for action: Recommendations for school mathematics of the 1980s.* Reston, VA: Author.

National Council of Teachers of Mathematics. (1989). *Curriculum and evaluation standards for school mathematics.* Reston, VA: Author.

National Council of Teachers of Mathematics. (1991). *Professional standards for teaching mathematics.* Reston, VA: Author.

National Research Council. (1989). *Everybody counts: A report to the nation on the future of mathematics education.* Washington, DC: National Academy Press.

National Research Council. (1990). *Reshaping school mathematics: A philosophy and framework for curriculum.* Washington, DC: National Academy Press.

National Research Council. (1991). *Moving beyond myths: Revitalizing undergraduate mathematics.* Washington, DC: National Academy Press.

Future Directions for the Study of Calculators in Mathematics Classrooms

Hersholt C. Waxman
Susan E. Williams
University of Houston

George W. Bright
University of North Carolina at Greensboro

In recent years, calculators have been described as having the potential to transform or change school mathematics (Burrill, 1992; Wheatley & Shumway, 1992). We certainly agree with that notion, but as we think about the information presented in the chapters in this volume, there are still a variety of concerns that remain with us. We need to look back at what we currently know about the use of calculators in teaching mathematics, and we need to reflect on how our current knowledge is different, both in quantity and quality, from what we knew a few years ago. Furthermore, we maintain that it is equally important to look into the future as best we can and determine not only what additional knowledge we need, but also how future developments in calculator technology might influence the knowledge we seek.

Much of the research and development effort that has been devoted to incorporating calculators into instruction has focused on creating model instructional materials (i.e., materials which reflect a coherent instructional philosophy and approach -- referred to as "models"), either for students directly or for their teachers (e.g., Bright, Lamphere, & Usnick, 1992). As would be expected, more recently developed models seem to have become increasingly successful at emphasizing the mathematics that can be taught with calculator technology rather than just the technology itself (e.g., keystrokes). We suspect that putting substantially more effort into development of additional models will begin to have decreasing payoffs. Rather, we argue that the next round of research and development effort should be devoted first, to careful evaluation of the effects of existing models on learning and second, to the study of ways to help teachers learn to use those models effectively.

We do, however, want to support many of the existing models that have been developed (e.g., Demana & Leitzel, 1988; Usiskin, 1987; Usiskin, Flanders, Hynes, Polonsky, Porter, & Victora, 1990). These models have been developed with great care and display considerable insight about both mathematics and the processes by which people learn that mathematics. The people who have developed these models should be acknowledged as the visionaries that they are. Without their efforts we would know considerably less about the ways that calculators can be used not only to improve instruction of familiar mathematics but also to structure instruction of unusual mathematics that is made accessible because of the power of calculators. It is precisely because of their successes that we can now move on to the next stage of learning about the impact of calculators on mathematics instruction, namely, what do students learn about mathematics when calculators are used regularly and powerfully in instruction.

Research Directions

In this section, we describe four specific areas that have important implications for future research in the field. First, we examine the impact of theory and research from the field of cognitive psychology on the use of calculators in mathematics classrooms. Second, we discuss needed research studies in the area of staff development and training. Third, we briefly look at issues related to evaluative research studies. Finally, we discuss the need to have more programmatic research in this area.

Cognitive Psychology Perspectives

Although there have been numerous studies of how calculator use affects student learning (c.f., Hembree & Dessart, 1986, 1992; Suydam, 1990), most of those studies have not systematically varied the ways that calculators were used with students. Rather, most studies have let teachers do what they will, and then tried to measure the effects on traditional mathematics tests. While this approach has some appeal as "action research," we maintain that in order to take full advantage of the power of calculators in learning mathematics, we probably need to know more about how various, specific uses of calculators (e.g., using the fraction simplification capability to learn about common factors of two integers) affect the ways that students internalize or generate mathematical knowledge.

Influenced by theory and research from the field of cognitive psychology, many mathematics educators and researchers have recently adopted an information-processing view of learning and teaching problem solving (Greeno, 1988; Resnick, 1987, 1988; Schoenfeld, 1988; Silver & Kilpatrick, 1988). This cognitive psychology perspective views learning as an active process and teaching as a means of facilitating active student mental processing (Gagne, 1985). In this paradigm, the focus changes from the student as passive recipient of information to the student as an active mediator of learning experiences who actively constructs or "generates" meaning from experience (Wittrock, 1974, 1978, 1986).

In mathematics education, the cognitive psychology perspective has approached understanding of learning through emphasis on the area of problem-solving strategies that students use when they solve problems. This approach implies that students need to apply cognitive strategies in order to learn (Winne, 1985). Consequently, there have been several studies that have investigated the cognitive strategies or goal-directed sequences of mental operations that students use to approach and solve problems (Gagne, 1985; Silver, 1985; Waxman, Padron, & Knight, 1991). Although the results of these studies have revealed the existence of cognitive strategies which enhance or detract from the problem solving process, the findings of these studies have not been integrated directly nor applied directly to research on the use of calculators in the classroom. We do not know, for example, whether calculators might facilitate the use of strong cognitive strategies (e.g., imaging or looking for a pattern) or diminish the use of weak strategies (e.g., using objects or guessing). In addition, we do not know to what extent calculator use allows students to develop new strategies or foster the use of multiple strategies.

Consequently, more research on the cognitive mathematics strategies students use with calculators is needed. This is especially true, because we know that calculators have a particularly beneficial effect on students' mathematical problem solving (Hembree & Dessart, 1992; Wheatley & Shumway, 1992). Research in this area is also needed to support the value of teaching students *when* and *how* to use these strong cognitive problem-solving strategies in conjunction with *when* and *how* to use calculators. Moreover, research on cognitive strategies needs to examine existing student strategies since they may create a situation which impedes student performance or prevents the acquisition of new strategies (Holley & Dansereau, 1984). In fact, finding ways to develop and extend students' "natural" pre-existing cognitive strategies may be more economical than attempting to develop new strategies for each situation, especially when calculators are initially introduced to students. We also need information on how calculator use affects existing differences. Finally, studies should specifically identify individual differences in strategy use as well as calculator use in order to see if there are cultural, sex-related, grade-related, or ability-related differences.

Staff Development and Training

Most of the participants in the Houston conference addressed the importance of providing teacher inservice on how to use calculators effectively. It seems clear that teachers who themselves did not learn mathematics in the presence of calculators (or any other electronic technology!) typically do not understand how best to take advantage of the power that calculators provide. Of course they can learn, but that learning requires considerable time and reflection. For example, Prokosch, Bright, and Freiberg (1991) reported that after a three-year inservice, middle school mathematics teachers' use of technology supported more time on task (97%) than has typically been reported for middle school mathematics classes. We believe that it is unlikely that teachers will be able to acquire the necessary knowledge and the confidence to act on that knowledge without systematic inservice programs. Consequently, we need to know a lot more about how effective inservice programs can be organized and delivered.

More research is also needed on preservice and inservice programs on calculators. Some approaches like the "training of trainers" model (Bright, Lamphere, & Usnick, 1992), "implementation support system" (Bitter & Hatfield, 1992), or the "coaching of teachers" (Copley & Williams, 1993) seem to hold promise, but more systematic studies are needed to examine the effectiveness of these and other approaches. In addition, experimental studies are needed that incorporate and test the simultaneous use of several of these training models and consequently provide information on the benefits and limitations of these and other approaches.

Along with the development of teacher inservice programs, we need to study the effects of those programs. Do teachers develop new understandings of the mathematics that they teach? Do teachers learn more about mathematics pedagogy? How does calculator use affect the interactions of students and teachers? How does it affect the classroom climate? Are effective inservice techniques also effective in preservice programs? The information provided by Copley et al. in this volume and elsewhere (e.g., Williams, Copley, Huang, & Bright, 1993) begin to establish norms of understanding about how the behavior of teachers changes when technology is injected into the classroom in the context of substantial teacher inservice. More information like this is clearly needed.

Another important issue related to the training of teachers is the attitudes of teachers and students toward calculators. Several studies, for example, have found that students and teachers fear that calculators will hinder understanding of basic computation skills (Bitter & Hatfield, 1993; Bright & Love, 1993; Huang, 1993; Huang, Copley, Williams, & Waxman, 1992). Consequently, the assessment of teacher and student attitudes is probably needed in most calculator projects and specific sessions or programs for dealing with negative attitudes towards calculators may have to be developed and implemented.

Evaluative Research Studies
Another important research area that still needs to be pursued is that of conducting better and more systematic, field-based, evaluative research studies. For example, it would be quite helpful to teachers and teacher educators to know how use of sets of coherent instructional materials, such as those developed by CAMP-LA (Pagni, 1991-92) affect what students learn. Evaluation of these sets of materials, in our minds, has been neglected. We do not know whether school districts (or even individual teachers) could be enticed into using such large sets of materials, but if evaluation data demonstrated important, significant positive effects, then we would at least have a reason to try to get those materials adopted in their entirety.

Another important aspect of evaluative studies that has often been neglected is the systematic observation of calculator use in mathematics classrooms. Many prior studies have simply examined the overall differences between the group that received calculators (i.e., experimental group) and the group that did not receive the calculators (i.e., the control group). The Copley et al. chapter in this book, however, illustrates the importance of observing how, and how often, calculators are used in the classroom. Other observational studies have similarly found that the use of calculators differs among classrooms as well as among individual students (Williams, et al., 1993; Huang & Waxman, 1993).

A final noteworthy issue related to evaluative studies concerns the assessment of student learning related to calculator use. Most prior studies have examined the impact of calculators on students' cognitive learning without either including calculator-sensitive items or allowing students to use calculators on the outcome measures. More work needs to be done on the impact of calculators when calculator-sensitive assessments are used to examine the effectiveness of programs.

Programmatic Research
There have been several exemplary studies in the field of calculators in mathematics classrooms, but more programmatic research is needed. Professional organizations like the National Council of Teachers of Mathematics and the Society for Technology and Teacher Education, may facilitate programmatic research through its conferences, journals, and monograph series. Other organizations like the National Science Foundation and the federally-funded national research centers on mathematics can also encourage concerted research efforts.

Although some of the research summarized by the articles in this book suggests several consistent relationships between calculators and teacher and student outcomes, further descriptive, correlational, longitudinal, and especially experimental research is needed to verify these results. In addition, studies should attempt to replicate some of the previous studies in other settings, especially in inner-city settings where many more students are at risk of dropping out and not furthering their education and where teachers are considered to be working in at-risk school environments (Waxman, 1992). Research incorporating contextual effects and individual differences is also important since prior research has suggested that there are differences by grade and students' ability (Hembree & Dessart, 1992). Furthermore, there have been very few studies that have addressed the impact of calculators on second language learners. This is an especially critical area given the growing numbers of second language learners in schools and since some research has already found that there are several significant differences on the extent to which limited-English proficient students use calculators in the classroom (Waxman, Huang, Williams, & Copley, 1992). These and other issues still need to be examined so that we can continue to understand how calculators influence teacher and student outcomes, as well as how calculators can change mathematics classrooms.

Future Considerations

It is more difficult to look into the future and project what we would like to know. If the calculators of ten years from now were exactly like the ones we have today, then we could with some confidence say things like, "We need to know more about how calculators' graphing capabilities affect students' understanding of function." These needs would be designed to complete the picture of our understanding of how calculators affect learning. Unfortunately, the calculators of ten years from now will not be the same as those we have today. One of the changes unanimously requested by participants at the conference in Houston was the merging of "large screen displays" such as are available on graphing calculators with the fraction capabilities found on the TI Math Explorer calculator. Most of us intuitively felt that this addition would make the fraction calculator much more effective as a teaching tool.

Today's world is a mix of good news and bad news. The good news is that technology seems likely to revitalize the process of curriculum more than any other change in the history of the school curriculum. There will be opportunities to improve instruction in ways that would not be available if technology were not driving the changes. The bad news is that technology is changing so rapidly that once we think we understand important relationships between the use of technology and the quality of the teachers and teaching, the technology will have changed dramatically. We'll be back in the position of needing to do additional research to understand new relationships. Research takes time to plan, conduct, and interpret. But the technology that is the focus of that research will continue to change, and the relationships we establish for one type of technology may not transfer to new technologies.

One possible response to this state of affairs is to become overwhelmed at the prospect of never reaching the end of the quest. However, the energy and enthusiasm of the people who have contributed to this volume are certain to act as an antidote to the lethargy that tends to result from feelings of being overwhelmed. Understanding the role of calculators in the K-12 mathematics curriculum is an enormous challenge. But we believe that our colleagues are not only qualified to meet that challenge but also dedicated to making calculators serve the purpose of improving mathematics instruction in our nations' schools.

References

Bitter, G. G., & Hatfield, M. M. (1992). Implementing calculators in a middle school mathematics program. In J. T. Fey & C. R. Hirsch (Eds.), *Calculators in mathematics education: 1992 yearbook* (pp. 200-207). Reston, VA: National Council of Teachers of Mathematics.

Bitter, G. G., & Hatfield, M. M. (1993). Integration of the math explorer calculator into the mathematics curriculum: The calculator project report. *Journal of Computers in Mathematics and Science Teaching, 12*(1) 59-81.

Bright, G. W., Lamphere, P., & Usnick, V. E. (1992). Statewide in-service programs on calculators in mathematics teaching. In J. T. Fey & C. R. Hirsch (Eds.), *Calculators in mathematics education: 1992 yearbook* (pp. 217-225). Reston, VA: National Council of Teachers of Mathematics.

Bright, G. W., & Love, W. P. (1993). Introductory calculator inservice for middle school mathematics teachers. In D. Carey, R. Carey, D. A. Willis, & J. Willis (Eds.), *Technology and teacher education annual 1993* (pp. 590-595). Charlottesville, VA: Association for the Advancement of Computing in Education.

Burrill, G. (1992). The graphing calculator: A tool for change. In J. T. Fey & C. R. Hirsch (Eds.), *Calculators in mathematics education: 1992 yearbook* (pp. 14-22). Reston, VA: National Council of Teachers of Mathematics.

Copley, J. V., & Williams, S. E. (1993). The effect of coaching on the use of technology in middle school mathematics classrooms. In D. Carey, R. Carey, D. A. Willis, & J. Willis (Eds.), *Technology and teacher education annual 1993* (pp. 51-54). Charlottesville, VA: Association for the Advancement of Computing in Education.

Demana, F., & Leitzel, J. R. (1988). Establishing fundamental concepts through numerical problem solving. In A. F. Coxford & A. P. Shulte (Eds.), *The ideas of algebra, K-12: 1988 yearbook* (pp. 61-88). Reston, VA: National Council of Teachers of Mathematics.

Hembree, R., & Dessart, D. J. (1986). Effects of hand-held calculators in precollege mathematics education: a meta-analysis. *Journal for Research in Mathematics Education, 17,* 83-99.

Hembree, R., & Dessart, D. J. (1992). Research on calculators in mathematics education. In J. T. Fey & C. R. Hirsch (Eds.), *Calculators in mathematics education: 1992 yearbook* (pp. 23-32). Reston, VA: National Council of Teachers of Mathematics.

Gagne, E. (1985). *The cognitive psychology of school learning.* Boston: Little Brown.

Greeno, J. G. (1988). For the study of mathematics epistemology. In R. Charles & E. Silver (Eds.), *The teaching and assessing of mathematics problem solving* (pp. 23-31). Reston, VA: National Council for Teachers of Mathematics.

Holley, C., & Dansereau, D. (1984). The development of spatial learning strategies. In C. Holley & D. Dansereau (Eds.), *Spatial learning strategies: Techniques, applications, and related issues* (pp. 189-209). New York: Academic Press.

Huang, S. L. (1993). Investigating middle school students' attitudes toward calculator use. In D. Carey, R. Carey, D. A. Willis, & J. Willis (Eds.), *Technology and teacher education annual 1993* (pp. 514-518). Charlottesville, VA: Association for the Advancement of Computing in Education.

Huang, S. L., Copley, J. V., Williams, S. E., & Waxman, H. C. (1992). Investigating middle school mathematics teachers' attitudes toward calculator use.

In D. Carey, R. Carey, D. A. Willis, & J. Willis (Eds.), *Technology and teacher education annual 1992* (pp. 401-405). Charlottesville, VA: Association for the Advancement of Computing in Education.

Huang, S. L., & Waxman, H. C. (1993). Classroom observations of middle school students' technology use in mathematics. In D. Carey, R. Carey, D. A. Willis, & J. Willis (Eds.), *Technology and teacher education annual 1993* (pp. 519-523). Charlottesville, VA: Association for the Advancement of Computing in Education.

Pagni, D. L. (1991-92). Calculator usage at the middle school level: A national survey. *Journal of Educational Technology Systems, 20*, 59-71.

Prokosch, N. E., Bright, G. W., & Freiberg, H. J. (1991, April). *Classroom observations in middle school mathematics taught with technology*. Paper presented at the annual meeting of the American Educational Research Association, Chicago.

Resnick, L. B. (1987). *Education and learning to think*. Washington, DC: National Academy Press.

Resnick, L. B. (1988). Treating mathematics as an ill-structured discipline. In R. Charles & E. Silver (Eds.), *The teaching and assessing of mathematics problem solving* (pp. 32-60). Reston, VA: National Council of Teachers of Mathematics.

Schoenfeld, A. H. (1988). Problem solving in context(s). In R. Charles & E. Silver (Eds.), *The teaching and assessing of mathematics problem solving* (pp. 82-92). Reston, VA: National Council of Teachers of Mathematics.

Silver, E. A. (1985). Research on teaching mathematical problem solving. In E. A. Silver (Ed.), *Problem solving and education: Issues in teaching and research* (pp. 81-96). Hillsdale, NJ: Lawrence Erlbaum.

Silver, E. A., & Kilpatrick, J. (1988). Testing mathematical problem solving. In R. Charles & E. Silver (Eds.), *The teaching and assessing of mathematics problem solving* (pp. 178-186). Reston, VA: National Council for Teachers of Mathematics.

Suydam, M. (1990). *Research on the use of calculators in mathematics instruction*. Unpublished paper.

Usiskin, Z. (1987). Lessons learned from the first eighteen months of the secondary component of the UCSMP. In I. Wirzup & R. Streit (Eds.), *Developments in school mathematics around the world* (pp. 418-429). Reston, VA: National Council of the Teachers of Mathematics.

Usiskin, Z., Flanders, J., Hynes, C., Polonsky, L., Porter, S., & Viktora, S. (1990). *Transition mathematics*. Glenview, IL: Scott, Foresman.

Waxman, H. C. (1992). Reversing the cycle of educational failure for students in at-risk school environments. In H. C. Waxman, J. Walker de Felix, J. Anderson, & H. P. Baptiste (Eds.), *Students at risk in at-risk schools: Improving environments for learning* (pp. 1-9). Newbury Park, CA: Corwin/Sage.

Waxman, H. C., Huang, S. L., Williams, S. E., & Copley, J. (1992, August). *The effect of calculator use in middle schools on limited English proficient students problem-solving achievement in mathematics*. Paper presented at the Third National Research Symposium on Limited English Proficient Students, Arlington, VA.

Waxman, H. C., Padron, Y. N., & Knight, S. L. (1991). Risks associated with students' limited cognitive mastery. In M. C. Wang, M. C. Reynolds, & H. J. Walberg (Eds.), *Handbook of special education: Emerging programs* (Volume 4, pp. 235-254). Oxford, England: Pergamon.

Wheatley, G. H., & Shumway, R. (1992). The potential for calculators to transform elementary school mathematics. In J. T. Fey & C. R. Hirsch (Eds.),

Calculators in mathematics education: 1992 yearbook (pp. 1-13). Reston, VA: National Council of Teachers of Mathematics.

Williams, S. E., Copley, J. V., Huang, S. L., & Bright, G. W. (1993). Effect of teacher involvement in curriculum development on the implementation of calculators. *Journal of Technology and Teacher Education, 1*(1), 53-62.

Winne, P. (1985). Steps toward promoting cognitive achievements. *The Elementary School Journal, 85*, 673-693.

Wittrock, M. (1974). Learning as a generative process. *Educational Psychologist, 13*, 15-29.

Wittrock, M. (1978). The cognitive movement in instruction. *Educational Psychologist, 11*, 87-95.

Wittrock, M. (1986). Students' thought processes. In M. Wittrock (Ed.), *Handbook of research on teaching* (3rd ed., pp. 214-229). New York: Macmillan.

www.ingramcontent.com/pod-product-compliance
Lightning Source LLC
Chambersburg PA
CBHW030757180526
45163CB00003B/1063